The Cherokee Physician, or Indian Guide to Health

The Cherokee Physician, or Indian Guide to Health

The Cherokee Physician, or Indian Guide to Health

The Cherokee Physician Or Indian Guide to Health, as Given by Richard Foreman, a Cherokee Doctor; Comprising a Brief View of Anatomy.: With General Rules for Preserving Health Without the Use of Medicine

[Special Illustrated Edition]

Authored by Richard Foreman, Authored by Jas. W. Mahoney

Illustrated By J. Mitchell

The Cherokee Physician, or Indian Guide to Health

The Cherokee Physician, or Indian Guide to Health

The Cherokee Physician Or Indian Guide to Health, as Given by Richard Foreman, a Cherokee Doctor; Comprising a Brief View of Anatomy.: With General Rules for Preserving Health Without the Use of Medicine

[Special Illustrated Edition]

Authored by Richard Foreman, Authored by Jas. W. Mahoney

Illustrated By J. Mitchell

©2011

J. Mitchell
ISBN: 978-1-105-26293-7

DISTRICT OF EAST TENNESSEE:

Be it Remembered, That on the 6th day of October, 1845, William Mahoney, of said District, hath deposited in this office, the Title of a Book, which is in the words and figures following, to wit:

"The Cherokee Physician, or Indian Guide to Health,
"as given by Richard Foreman, a Cherokee Doctor, com-
"prising a brief view of Anatomy, with General Rules
"for Preserving Health, without the use of Medicines.--
"It also contains a description of a variety of Herbs and
"Roots, many of which are not explained in any other
"Book, and their medical virtues have hitherto been un-
"known to the Whites; to which is added a short Dispen-
"satory, by James W. Mahoney," the right whereof he claims as Proprietor, in conformity with an act of Congress, entitled an act to amend the several acts respecting copy-rights.

A true copy from the Records in my office.

SEAL.

JAMES W. CALDWELL,
*Clerk of the United States
Court for the District of
East Tennessee.*

The Cherokee Physician, or Indian Guide to Health, as Given by Richard Foreman, a Cherokee Doctor; Comprising a Brief View of Anatomy, With General Rules for Preserving Health without the Use of Medicines. The Diseases of the U. States, with Their Symptoms, Causes, and Means of Prevention, are Treated on in a Satisfactory Manner. It Also Contains a Description of a Variety of Herbs and Roots, Many of which are not Explained in Any Other Book, and their Medical Virtues have Hitherto been Unknown to the Whites; To which is Added a Short Dispensatory

.INTRODUCTION.

Every new publication on this, as well as other subjects, should have some grounds upon which it can set up its claim to a share of public patronage and support. Had I not believed that this work contained something new and useful, I would not have published it. But, believing as I do, that the "HEALING ART," as known and practised by the Cherokee Indians, would be welcomely received by many: and having personally tested the efficacy of their remedies, in the cure of diseases after such remedies as are usually prescribed by the whites had been tried and had utterly failed to effect a cure, I have been induced to commit this system to paper. I am sensible that in so doing, I expose myself to the animadversions of the critics. I am also sensible, (to some extent) of the prejudice which prevails in the minds of many, against Medical works, which are not decked in the flowery drapery of a fine and ornamented style and technical lore. Believing, as I do, that medicine should not be merely a study of curious enquiry, but one of the deepest interest to every son of mortality, I have endeavored to adorn it with plain practical sense, rather than with the fascinating decorations of high standing, unmeaning names, and technical phrases.

Those who will take the pains to read and study, will soon be convinced that the All-wise Creator in the infinitude of his mercy, has furnished man with the means of curing his own diseases, in all the climates and countries of which he is an inhabitant; and that a knowledge of the means of curing all common diseases, is not so difficult to obtain as has been generally represented.

The really valuable materials in medicine, and those which act with the greatest promptitude and power, in the cure of diseases, are few and simple, and easily to be procured in all countries.

The Aborigines of our country found the means of mitigating and curing their diseases, in the uncultivated wilds which gave them birth,--they knew nothing of foreign drugs, but with roots, herbs, and plants found in their own country, they mitigated and cured the diseases most common to that country. That their knowledge of the medical properties of the roots and herbs common in the American forest, is superior to that possessed by the whites will hardly be denied. Neither will it be denied by those acquainted with their success, in treating disease, that they have, in many instances, performed cures, by means of roots, herbs and plants, after the usual remedies prescribed by white physicians had failed. The articles employed by them in the cure of diseases, are simple, and principally such as can be procured in this country.

The time is not far distant, when most; if not all the diseases, of our country, will be healed without the use of calomel and mercurial preparations, and when foreign drugs will be disused by administering physicians.

My principal design, in the publication of this work, is to lay before the heads of families, the means of guarding against diseases, and also such remedies as are best calculated to arrest diseases in their incipient, or forming stages. I have labored to give such instruction, with regard to the nature and symptoms of diseases, as will enable the reader to determine, with some degree of accuracy, when the aid of a skillful physician is really necessary, and also to distinguish the man of practical science and wisdom, from the ignorant pretender, and the assuming quack.

With these remarks, I submit the work to the inspection of a liberal and enlightened American people. The impartial and intelligent reader will doubtless award to it its due portion of merit and demerit.

The Cherokee Physician, or Indian Guide to Health

The Cherokee Physician, or Indian Guide to Health

PART ONE.

CHAPTER 1.

ANATOMY.

Anatomy treats of the structure of the human body, its various organs, and their use.

Practical Anatomy, is the dissecting or dividing of the organized substances, to exhibit the structure, situation, and uses of the parts. Those wishing to practice surgery, will find that subject discussed at length in books that treat on that alone. A knowledge of Anatomy is indispensable to him who would become either a safe or a skillful Surgeon; but to a practical Physician, in the treatment of diseases, it is of little value, comparatively speaking. But as this work is designed for all who may see proper to give it a perusal, and not limited to the use of any in particular, it is reasonable to suppose that some will be pleased, and perhaps benefitted, by this part of the work. A minute and extensive treatise on Anatomy will not be expected by the intelligent reader, in a work of this kind. But I will endeavor to give the outlines of the whole human system, in a plain and concise manner. This short treatise on this subject, will be sufficient to enable the heads of families, and others, who practice under the directions of this book, to ascertain with some degree of accuracy, the seat of disease, and also to enable them to return, to its proper place, a dislocated joint; and this is all that the writer believes will be worth its room in this work.

SECTION 1.
ORGANS OF THE HUMAN BODY & THEIR USES.

The most natural general divisions of the human body, are--

- 1. The head (*Craneum.*)
- 2. The body, (*Trunk.*)
- 3. The legs, feet and hands, (*upper and lower extremities.*)

These general divisions are composed of bones, muscles, glands, ligaments, cartilages, tendons, nerves, blood vessels, absorbents, and the brain and spinal marrow.

SUB-DIVISIONS.--The body, (*Trunk,*) is divided into two cavities:

- 1. The breast, (*Chest or Thorax.*)
- 2. The belly, (*Abdomen.*)

The breast, (*thorax*) and belly (*abdomen,*) are separated by a strong membrane, called the *midriff* or *diaphragm,* which will be described hereafter.

The upper division, breast (*thorax,*) contains the heart and lungs, called the *thoracic viscera;* and the lower division, belly (*abdomen*) contains the stomach, kidneys, liver, intestines, &c., called *abdominal viscera.*

The bones will now be taken into view. They may properly be considered as the braces of the human frame --they give to it shape, stature and firmness. The number of bones in the human body, is estimated at two hundred and forty-eight. Of these, sixty-three are in the head; fifty-three in the trunk; sixty-eight in the upper extremities, or arms, and sixty-four in the lower extremities.--This estimation includes the four *sesamoid* bones in the great toes, and the four *sceamoid* bones in the thumbs, which are not always found.

SKULL--(Cranium.)

The skull contains the eight following bones:

- One in the forehead--*os frontis.*
- Two temple bones--*ossa temporalia.*
- Two walls, or sides--*ossa parietalia.*
- One full of holes--*os ethmoides.*
- One wedge-like form--*os spenoides.*
- One back of the head--*os occipitis.*

The *os frontis,* is the bone of the forehead, reaching from its upper edge, downwards, so as to include the upper part of the eye sockets, and backwards on each side, so as to join the temple bones. The temple bones join the walls, or sides, and the forehead.

The *Os Ethmoides,* or bone full of holes, is a very curious bone, situated on the inside of the head, or rather forehead. It is a light spongy bone, having somewhat the appearance of net-work.

The *Os spenoides,* or bone of wedge-like form, spreads across the inside of the head, and attaches itself to fourteen other bones.

The *Os occipitis,* is the hind part of the head, and joins the neck bone; it is a very thick but uneven bone. It supports the hind part of the brain, and through it passes the marrow of the neck and back, called the *spinal marrow.* All the preceding bones are joined together by seams, which in appearance resemble saw-teeth.

The face is next in order, in which are many small bones. It has six bones on each side; and they all have seams similar to those of the skull, only smaller.

The nose bones, *ossa nasi,* are the two bones which form the nose, and meet together by two thin edges, without any indentations.

The upper jaw bones, *ossa maxillaria superiora,* which are large, and form the basis of the face. They extend upwards, and form the side of the nose, and they send backward a kind of plate, that makes the roof of the mouth. A circular projection below, makes the sockets for the teeth.

The Vomer, *a plough-shear,* completes the nose.

The cheek bone, *os malae,* is the high bone that forms the cheek.

The lower jaw-bone, *os maxillae inferioris,* has but two joints, those under each ear.

The spine, or back-bone, comes next in order. This is a long line of bones, extending from the back of the head to the end of the body. It has twenty bones, or joints, called *vertebra.* The neck part has seven joints, *vertebra;* the back twelve, and the loins five; making in all, twenty-four separate bones. In some persons the neck has eight pieces, the back eleven, and the loins six. Some persons, with very short necks, only have five pieces in the neck, and the number made up in the loins. The same marrow runs from the back of the head to the lower end of the spine.

SHOULDER BLADE--(*Scapula.*)--

The shape and situation of this bone is so well known, that it needs no explanation. It is not connected to the trunk by ligaments, but has several muscular substances between it and the trunk.

COLLAR BONE--(*Clavsile.*)--

This is perhaps the strongest bone in the system, to its size. It is placed at the lower part of the neck, and reaches from the upper part of the breast bone to the point of the shoulder. It is fastened

by grisly substances--*cartilages,* and rolls with ease on any exertion of the breast and shoulder.

UPPER BONE OF THE ARM--(*Os Humeri.*)--

This bone has a cylindric form, but at the lower end it is twisted and flattened a little. This flatness joint it to the elbow in a hinge-like form, so that the joins has but one direction of moving. At the shoulder it has a large round head, which enables it to turn in every direction. On the top of the head, this bone, though circular, is nearly flat, and has but a very shallow cavity to turn in; consequently it is a very weak joint to its size, and easily dislocated.

LOWER PART OF THE ARM--(*Radius and Ulna.*)--

The lower part of the arm, from the elbow to the wrist, has two bones in it. The main bone has its largest end downwards, joining the wrist next to the thumb, while the little end is upwards, lying on the ulna, where the ulna joins the large bone of the arm at the elbow. The radius gives all turning motions to the wrists. It is a stronger bone than the ulna, and is somewhat arched in its shape. The upper end of this bone is small, of a button like shape, and is joined both with the large bone and the ulna. This bone gives more strength than the ulna to that part of the arm, particularly to the wrist.

THE ULNA OR ELBOW--(*a measure.*)

By this bone we perform all the actions of bending and extension. It is of a triangular form, and is so firmly attached to the upper bone of the arm, (os humeri) that it allows no lateral or side motion.

BONES OF THE HAND AND FINGERS.--

The wrist bones are eight in number. They are situated between the end of the arm bones and the bones of the hand: they are very short, and are bound together very strongly, by cross ligaments, and closely compressed together, so as to form a ball-like figure, each having separate ends or joints: there are five bones between the wrist and fingers--they start out from the wrist, each one extends to its finger respectfully: they are all nearly straight round bones, without joints, tolerably large and very strong: the fingers all have three joints, the thumb has two.

THE BREAST BONE--(*Sternum.*)--

This bone lies exactly

in the front part of the breast. It is a light spongy bone. In children, and in some to the age of five or six years old, this bone consists of eight distinct pieces, which in old persons become one solid bone: they are a little hollowed at the upper end, and on each upper corner, it has a joining or articulating hollow, at which place the ends of the collar bones are fastened by strong ligaments. Each side of this bone is so formed, as to receive all the ends of the ribs on their respective sides.

THE RIBS.--

There are twelve ribs on each side of the breast or chest, corresponding in number with the vertebra, or joints, in that part of the spine, or back-bone. Seven are called line ribs, because they join the breast-bone: the other five are vulgarly called short ribs, but by anatomists false ribs, because they do not join the breast-bone: the ribs are connected with the breast-bone with cartilages, and to the back-bone by joints.

BONES BELONGING TO THE BASIN--(*Pelvis.*)--

This part is formed of very strong, firm bones, standing in a kind of arch between the main trunk and the lower extremities. Each bone is large, and affords large strong sockets for the thigh bones. In grown persons it contains four bones: --the *Os sacrum,* the *Os coccygis,* and the two *ossa innominata.*

The *Os sacrum* and the *Os coccygis,* is called the false spine, or column, the point of them runs downwards, and the largest part is upwards. It runs along that part of the system vulgarly called the *rump. Os coccygis* (cuckoo's bill,) is the lower end of the back-bone. It tapers from the *Os sacrum,* or rump bone, to its termination, so as to form a sharp point. It is a little crooked and flattish, so as to support the lower gut, (*rectum*) bladder, and womb: it is very flexible, and recedes in time of labor with women, so as greatly to facilitate the passage of the child's head; and when labor is over, it returns to its proper position without difficulty.

The two *Ossa innominata,* or nameless bones, are two great bones that make the two sides of the basin, or pelvis. The *Os Ilium* is the greatest part of these bones. It extends up in a sort of wing from the pelvis, or basin, and is covered with the muscles that move the thighs.

The hip-bone, (*Os ischium*) lies directly under the flank bone, (*Os ilium*) and is the lowest point of the basin, or pelvis, vulgarly called the buttock, being the point on which we sit.

The share-bone, (*Os pubis,*) is the smallest piece belonging to the nameless bones, (*Ossa innominata.*) It completes the front part of the brim of the basin (pelvis.)

THIGH BONE--*Os Femoris.*--

This is the largest, longest, and most cylindrical bone belonging to the human anatomy. It joins the hip in a way that gives it strength. It is very hard to dislocate, or put in place. It has a regular bend from nearly one end to the other; the bending side being towards the front of the thigh:--this is the strongest joint in the body.

The leg bones, two in number, called by anatomists *Tibia and Fibula.* The tibia is the largest of the two leg bones, and is situated on the inside part of the leg. It is of a triangular form, with the upper end somewhat flattened: the fibula is on the outside of the tibia, and makes the outward lump of the ankle.

THE KNEE PAN--(*Rotella* or *Patella.*--

Is a small roundish bone, tolerably thick: it is attached to the tubercle of the tibia by very strong ligaments.

INSTEP OR ANKLE--(*Tarsus.*)--

The ankle is composed of seven bones, which lie between the leg and foot. They are bound together by ligaments, in a manner similar to those of the wrist. One of them forms the heel, and is called the heel bone--(*Os calcis.*) There are five bones between the ankles and toes: they join the ankle and toes in a similar manner to the hand bones.

I have now described the shape and position of such bones as are most liable to dislocation and injury: the next subject will be the internal parts of the human system.

THE BRAIN.

The brain is the great sensorium of the system, and has a communication through the nerves with the whole body. It receives all impressions made upon any of the organs of sense, and is really the seat of sensation. It is here that all the impressions made upon the organs of sense, are manufactured into ideas. But in what manner the brain performs this, or what connection it has with the mind, is a mystery in which the researches of physiologists, and the deductions of metaphysicians, have hitherto been unable to reflect any light. "The most, or in fact all that is known on the subject is, that the mind acquires all its ideas of external objects through impressions made by these objects on the organs of sense. These impressions are conveyed to the brain by the nerves, and produce what is called sensation, which is the passive reception of the image of the archetype, or pattern of the idea upon the brain, and in some unknown manner, the perception is conveyed to the mind."

The brain is situated in the upper cavity of the head. It is divided into two grand divisions, which are called:

- 1. The seat of imagination--*cerebrum.*
- 2. The seat of animal spirits--*cerebellum.*

There are several other smaller divisions.

"The brain is larger in man than any other known animal. Its general weight is from two pounds five and a half ounces, to three pounds three and three-quarter ounces: many however, weigh four pounds. The brain of Lord Byron (without its membranes) weighed 6 pounds."

The spinal marrow is only a continuation of the substance of the brain, through the cavity of the spine or back-bone.

THE TONGUE--(*Lingua.*)--

The tongue is composed of small muscular fibres; it is coursed with little reddish pimples, which are the ends, or terminations of nerves, it is the impression made on these nerves that produces that pleasurable sensation called taste.

THE WIND-PIPE--(*Trachea.*)--

This is a rough canal, through which the air passes from the mouth to the lights, (lungs) in breathing. It lies in front of the swallow. (*esophagus*) and every thing taken into the stomach, passes directly over the mouth of the wind-pipe; but it has a kind of lid or valve, that shuts or closes over it in the act of swallowing, (*deglutition.*) At or near the lungs it forks, or branches off, so as to convey the air into the lungs.

THE LIGHTS--(*Lungs.*)--

The lungs are situated in the chest, *thorax*). The thorax, or chest, is lined with a smooth shining membrane, denominated the *pleura,* which is the seat of, and gives name to the Pleurisy. The pleura forms two distinct apartments in the chest, two sides of which meeting, attach to the inner edge of the spine, or back-bone, and reaching from thence to the breast bone, form the partition called the *mediastinum.* The lungs are divided into two lobes or portions, and situated one in the right, and the other in the left side of the breast, in the above named apartments. They join the wind-pipe, *trachea,* in the upper part of the breast. They are attached to the heart by the pulmonary vessels. They are full of little tubes, which communicate with the external atmosphere through the wind-pipe.

The most important, and perhaps the only function of the lungs is that of breathing, *respiration,* which is simply inhaling the air into the lungs, and expelling it from them.

THE HEART.--

The heart is situated in the chest, or thorax, near the centre of the human body, with its main base placed a little on the right of the back-bone, and its point standing obliquely to the sixth rib, on the left side. As it lies in this oblique position, its under side or surface, is in contact with the diaphragm. It is so placed between the arteries and veins, as to regulate their relative action, in propelling the blood through the arteries, and receiving it through the veins. It is divided into two cavities, which are distinguished by the names of right and left ventricles. There are two other hollow muscles denominated auricles: the heart possesses the power of dilating and contracting, which is technically denominated the *systole* and *diastole* motion. By this operation it first receives the venous blood into its cavities, and then forces it into the arteries, by which it is carried to every part of the body. This motion continues day and night, awake or asleep, during the whole period of our lives. The number of these motions in a given period, is modified or governed by age, or by disease. In infancy the number is greatest, being from one hundred and thirty, to one hundred and forty; in manhood, from seventy to eighty; in old age, from fifty-five to sixty-five, in a minute. Most inflammatory diseases stimulate the muscles of the heart, and accelerate its motion. It is this power that rolls the "precious fluid" of life thro' every channel in the system, with the constancy of a perennial fountain. "While the vital spark remains, the HEART with untiring assiduity, plies the wheels of life, unfatigued with its ceaseless labor; and is neither lulled into stupidity by the torpor of sleep, nor decoyed into remissness by the enchantment of pleasure." It performs two circulations at the same time; that with the lungs, and that with the body. From the lungs it receives nothing but pure blood, and to the body it sends out such as is fit for its support.

THE SWALLOW--[*Esophagus.*]

This is a canal or tube, commencing at the mouth, and running downwards to the stomach, which it joins, and into which it empties the food. It lies close to the back-bone, behind the wind-pipe, and passes through the diaphragm.

The Diaphragm, or midriff, is a muscular substance, composed of two muscles; the upper one of which originates at the breast bone, and at the ends of the last ribs on each side: the second muscle starts at the back-bone of the loins; it is covered on its under side by the *peritoneum,* and on the upper side with the pleura. The gullet, great run, and several other vessels, pass through the diaphragm.

THE LIVER--(*Haper.*)--

The liver is situated immediately below the diaphragm to which it is attached. It is the largest organ in the system: it is divided into two principal lobes, the right of which is much the largest: the liver is connected with the gall bladder, (bile,) and billiary vessels; its office appears to be that of secreting the bile from the blood, which is necessary in the digestion of food. A portion of the bile is regularly thrown through the vessels of the liver and gall bladder, into the stomach.

The gall bladder, *vesicula fellis,* is attached to the liver, and lies in a cavity of the liver, on the underside. It is of an oblong form, and appears to be for the purpose of containing the bile, until the proper time for it to be thrown into the stomach:--the bile is conveyed from the gall bladder into the first portion of the small intestines, called *duodenum,* and from thence into the stomach.

THE STOMACH--[*Stomachus.*]--

The stomach is a large membranous substance, of an oblong, bag-like shape. Its most important use is to receive the masticated food, and retain it until the process of digestion is so far completed as to reduce the food to a pulpy, semi-fluid mass, called chyme. When digestion is so far advanced as to convert the food into chyme, [pronounced *kime,*] it is poured into the *duodenum,* where it mixes with the *pancreatic juice.* From this mass, the absorbent vessels, called *lacteals,* obtain a white opaque fluid termed chyle [pronounced *kile.*] Digestion is principally effected by the solvent powers of the gastric juice, which is a fluid secreted in the stomach. The solution of the food by the gastric juice, is supposed to be a chemical process decomposing it, and separating it into its elementary principles. The stomach may justly be considered one of the most important organs in the animal economy.

THE MELT--(*Spleen.*)--

This is not a vital part, as the other organs are, which have just been described. It has been removed from both man and beast, without the least apparent injury. It is attached to the stomach, and lies mostly in the left side.

The caul fat (*omentum*)

is situated under the membrane--*peritoneum,* that lines the belly, and above the intestines, it is a white gauzy looking substance, it assists in forming the bile, serves to guard the internal parts against cold--lubricates and softens such parts as are connected with it, and in a state of starvation it supports the system. This is one reason why a fat animal can sustain life so long without food. It is very beautiful, and rather singular in its appearance; it resembles a white piece of fine net work, which had been carelessly tossed down in a half folded position.

The pancreas --

a flesh organ--is situated under the stomach. It is of an irregular oblong form, by some compared to a dog's tongue. It is composed of glands, veins, nerves and little ducts or vessels, also something of a fleshy consistence. Its use appears to be that of secreting the juice that is to be mixed with the chyle.

INTESTINES OR GUTS--[*Intestinum.*]--

The intestines comprehend the whole tube, from the stomach to the fundament; their office is to receive all the food--retain it according to the laws of nature, and then pass off the crude or excrementitious part, according to the same.

THE KIDNEYS.--

The kidneys are situated outside of the lining of the belly, near the back-bone, and on each side: they are of a dull red color: it is their province to secrete the urine from the blood. Each kidney receives a large artery, which proceeds immediately from the *darta;* and a vein issuing from each kidney, returns the blood to the *vena cava,* after its superabundance of water has been separated from it. The urine is first secreted or collected in them, and excreted or thrown out, through the two canals called ureters, into the bladder. The ureters are about the size of a small goose quill. The kidneys are subject to derangement in the performance of their office, in two ways: First, the secretion may be checked, and a proper quantity of fluid not be carried off: and secondly, its secretion may be too active, and carry off too much of the fluids.

THE BLADDER--(*Vesica Urinari Cystis.*)--

The water bladder lies in the front part of the abdomen, within the basin. Its office is to receive the water, or urine, which is collected in the kidneys, through the ureters; the urine is next discharged by the neck of the bladder, through the urinary canal (urethra) which reaches from the neck of the bladder, to the end of the privates. The muscles at the neck of the bladder are possessed of very strong contractive powers, by which the bladder is enabled to retain the urine the natural length of time.

NERVES.--

The nerves are small white fibers: they all have their origin in the brain and spinal marrow: those which issue from the brain, are called *cerebral,* and are the organs of sensation: it is their province to convey impressions to the brain from all parts of the system:--those issuing from the marrow of the spine, are termed spinal; it is their province to communicate the power of motion to the muscles. The nerves all issue in pairs: there are usually reckoned forty pair of nerves, nine of which have their origin in the brain, and thirty-one in the spinal marrow. It is by means of those that issue from the brain, that we hear, see, taste smell and feel; or in other words, they convey to the brain, the impressions received by the five organs of sense, in the act of seeing; feeling, hearing, tasting and smelling. A chord of nerves accompanies every

artery tolerably close. It is supposed that each fiber of the nerves, is a canal or tube, through which the nervous fluids pass, and communicate with each other, similar to the blood vessels.

THE ARTERIES--(*Arteria.*)--

The arteries are two in number: First, the great artery, *dorta;* second, the artery of the lungs, (pulmonary artery.) The great artery *dorta,* originates at the left ventricle or cavity of the heart, and is the greatest blood vessel in the body:--the pulmonary artery starts from the right cavity of the heart; all others are nothing more than branches of these:--the blood is thrown out from the heart, through the arteries, to every part of the body. As the blood passes through the arteries, the absorbent vessels of every part of the system, receive their respective portions of the nutritious properties of the blood. The arteries gradually become smaller as they proceed from the heart, and terminate in the veins through the capillary vessels: these little vessels connect the arteries and veins:-- the arteries are susceptible of considerable dilation and elongation, which takes place when the blood is forced into the dorta by the contraction of the heart, and when the action of the heart ceases, the effort of the artery to return to its usual dimension, keeps a constant motion of the blood along the arteries, during the dilation of the heart to receive another portion of blood, which by the contraction of the heart, is again driven into the *dorta,* and thus the vital tide is kept in motion. The dorta has a valve at its orifice, or opening into the heart, which readily admits the passage of the blood from the heart into the artery, but prevents its return from the artery into the heart:--the blood when it leaves the heart is of a bright red color, but as it returns through the veins to the heart, is of a dark purple color.

THE VEINS.--

All have their origin or commencement, at the ends of the arteries as I before stated:--the veins as they proceed from the extremities toward the heart, become larger by numerous branches intercepting each other and uniting, until they are all concentrated in two canals, termed *vena cava.* The veins have no pulsation as the arteries have; but in them the blood moves smoothly and slowly on; it is forced through the veins by a contractile power which they possess; and as the blood has mostly to run upwards in the veins, they are supplied with little valves, similar to those of a force pump, so that as the blood ascends in the veins, the lid (valve) gives way till the blood passes, then shuts or closes the place, so that no blood can fall back. The blood in passing through the lungs undergoes a great change; when it enters the lungs it is of a dark purple color, but when it leaves them and returns to the heart, it is of a bright red color: this change is produced by the air inhaled into the lungs. The blood in passing through the numerous delicate vessels in the lungs, absorbs oxygen from the air; and the air abstracts carbon from the blood. When the air is exhaled from the lungs, a great portion of its oxygen has disappeared, and carbon is found in its place:--the blood supplied with oxygen and relieved from its superabundance of carbon, is essentially revived, and sets out again, to distribute its fresh supply of nutrition and stimulus, to the different parts of the system.

THE MUSCLES.--

These serve to perfect the form and complete the symmetry of the body, but their most important use is to act upon the bones and produce animal motion: they terminate at the ends in grisly substances, by which they are fastened to the various parts of the system. Each muscle consists of a distinct portion of flesh, and has the power of contraction and relaxation: they are all in pairs except nine: there are reckoned one hundred and ninety-eight pair in the human system; this estimation makes the number of muscles four hundred and five.

THE GLANDS.--

The glands are composed of blood vessels nerves, and absorbents. They are distinguished according to the nature of their fluid contents, into mucous, sebaceous, lymphatic, lacrymal and salival glands.

The mucous glands are situated in the nose, back part of the mouth, throat, stomach, intestines, bladder, &c, and secrete (which means to separate from the blood) mucous, for the purpose of moistening all the internal surfaces that need moisture.

The sebaceous glands are situated in the arm-pits, face pubes, &c.: they secrete an oily substance.

The lymphatic glands are situated in the arm-pits, mesentary, groin, &c.

The salival glands are situated about the root of the tongue and angle of the jaw: they secrete the substance called saliva or spittle, which is discharged into the mouth.

The lacrymal glands are situated above the outer corners of the eyes: they secrete the fluid called tears, which serves to moisten the eyes and aid in expelling any extraneous matter from them. Grief and sometimes joy, operates in some unknown manner on the lachrymal glands, so as to produce a copious flow of the lachryma, or tears.

BREASTS OF FEMALES--[*Mammæ.*]--

The breasts of females are also regarded as glandular bodies: they are composed of a vast number of small ducts or vessels, which secrete the milk from the blood. The vessels which secrete the milk as they approach the nipple, fall into each other and form eight or ten large tubes, which are so admirably connected, that if anything obstructs the passage of the milk through one of these, it is discharged through the others without inconvenience.

JOINTS, GRISTLES--[*Cartilages.*]--

The joints (articulations) are fastened together with white gristly substances called cartilages: they are of the same texture and nature of the sinews and tendons: they are very strong and lasting.

JOINT WATER--(*Synovia.*)--

This is a new kind of oily substance that is contained in the joints, for the purpose of lubricating them: it greatly facilitates their motion; but if this juice or synovial water be extracted or discharged, by a cut or otherwise, it never can be restored, but the joint will remain stiff.

THE SINEWS--(*Tendons.*)--

By anatomists, the sinews or leaders, are called the terminations or extremities of the muscles. They are white gristly substances, very strong, and may be split into the finest threads imaginable.--They are very nearly the same in the human system as in animals. They were employed by the aboriginees of our country, in making moccasins, belts, &c., after splitting them to the proper size.

CHAPTER TWO.

The art of preserving Health without the use of MEDICINES.

The enjoyment of perfect health, is certainly one of the greatest earthly blessings that falls to the lot of mortals. Without health, honor, title, wealth, beauty, the kindness of friendship and the tenderness of affection, are all insufficient to render man even comfortable. All these blessings fail to relieve the pangs of disease, and give a relish to the affairs of life. The vast importance of health will render a short treatise on its preservation an acceptable article in this work. It will doubtless be readily acknowledged by all, that it is much better to shun or avoid disease, than to remove or overcome it after it has once taken hold on the system; and as the greatest number of our diseases and infirmities are the fruits of infringements on healthy laws of nature, how earnestly should we be engaged in correcting and avoiding those infringements. Man, in the early days of nature, lived in a state of perfect health, both in body and in mind. The friendly hand of nature gave him sustenance, without labor or toil, and nature's beverage quenched his thirst without the aid of spirituous liquors. Protected by the immediate presence of the Almighty, innocent of any violation of his law--living in the full enjoyment of his benevolence, man was happy. But alas! we now view him in a fallen state: he has transgressed the sacred laws of his Creator, God, and incurred the penalties annexed to his transgression. "His days are shortened and encumbered with disease." What a solemn thought, and how anxiously engaged should we be to change our condition; and how careful should we be to guard against evil by a temperate course in all things.--Health can only be secured and retained by temperate habits; it is a jewel, generally found in the possession of those only who have "moral firmness enough to curb their lust, check their appetites, control their passions, and submit to the regulations of virtuous and temperate habits.--Irregularity and intemperance in eating, drinking, sleeping and exercise, lays the foundation of most diseases with which the human family is afflicted.

> "Would you extend your narrow span,
> And make the most of life you can;
> Would you when medicines cannot save,
> Descend with ease into the grave?
> Calmly retire like evening light!
> And cheerful bid the world good-night,
> Let virtue and temperance preside.
> Our best physician, friend & guide."

SECTION I.
OF AIR.

Much might be said relative to the different gasses which compose the atmosphere, or air; for it is not as many persons suppose, a simple element, but is composed of unequal portions of oxygen, nitrogen, and carbonic acid. But as a scientific treatise on this subject properly belongs to the chemist, I will leave the subject with him, and confine my observations more particularly to the effects which the different states of the atmosphere has on the body.

Air is rendered impure and unwholesome in many ways; such air should be avoided as much as practicable. The air in cities, crowded assemblies, whether in-doors or out, is not wholesome. That in deep wells, damp cellars, close dungeons, caves, &c., is apt to become infected. Many persons have instantly expired on going down into deep wells or caves, where air composed of undue proportions of the above named gasses had settled. It may readily be ascertained whether a well or cave contains such air, by putting in them a lighted candle. If the candle continues to burn, the air is composed of such proportions of the different gasses, as is necessary for the support of animal life, and may be entered with safety; but if the candle goes out, the air is not such as will support animal life, but will produce instantaneous death. Air confined in close apartments where there is hot fires, is pernicious to health. Many persons injure their health by sitting or lying in rooms kept hot by large fires and not sufficiently ventilated, or dried. Air extremely hot or cold, is equally avoided, and should be equally avoided, particularly by persons of delicate constitutions. Night air is very pernicious to health, as is also the air between sunset and dark.

The body may be comfortably clad, and yet much injury to the health be sustained by exposure to a damp cold atmosphere; for it should always be remembered that it is equally dangerous to inhale it into the lungs, as to admit its free access to the external surface of the body; the consumptive and asthmatic should bear this well in mind, if they would value their own safety.

Dry air moderately cool, is the most salubrious bath to the healthy and infirm. A strong current of air should always be avoided. Never sit or lie in a current of air passing through a window or door, especially while warm; it checks perspiration, chills the blood, and often lays the foundation of incurable diseases.

SECTION II.
EXERCISES.--

Moderate and regular exercise is as essential to the preservation of health, as food is to the support of our bodies. It keeps up a regular circulation of the fluids, aids digestion, promotes the necessary secretions and excretions, and invigorates the frame. It prepares the body to be refreshed with sleep, and makes even the bed of straw or the hunter's blanket pleasant: "it furnishes an appetite that relishes plain and wholesome food, and preserves the healthy tone of the digestive organs. It gives clearness to the brain, vivacity to the spirits, cheerfulness to the mind, and elasticity to the whole system."

Exercise increases the strength of our nerves, of our muscles, of our sinews, and invigorates every fiber of the whole system. To prove this, we have only to turn our attention to the aboriginees of America. They spent their lives in the active pursuits of the chase in the open air; their diet and dress were of the simplest kind; they rose from their blankets at the dawn of morn, after having enjoyed a refreshing nights sleep, and prepared themselves for their homely but wholesome repast by active exercise in the open air. A knowledge of their habits, lives, diseases, &c., will also show that exercise is a great guarantee against a host of diseases with which the "pale-face" is so often afflicted, but is seldom found in the wig-wams of the "red man." Among these are Consumption, liver complaints, dyspepsy, hysterics, and many others too tedious to mention. Exercise is necessary from infancy. Only look at country children, who are accustomed to exercise and industry, how much more active and stout they are, than those of large towns, where they are cooped up in small rooms. Also look at the rich and indolent, and those who labor for their living. While the opulent and idle complain of ill-feeling and nervous weakness, the man of moderate exercise is vigorous, his appetite good, his sleep refreshing, and his mind cheerful. More than half of the female diseases, especially such as are connected with hysterics and nervous affections, arise from want of due exercise in the open and pure air.

> "How sweet at early dawn to rise,
> And view the glories of the skies;
> To mark with curious eye the sun,
> Begin his radiant course to run;
> Her fairest form then nature wears,
> And clad in brightest green appears.
> Nor you, ye delicate and fair,
> Neglect to taste the morning air,
> It will your nerves with vigor brace.
> Improve and heighten every grace;
> Add to your breath a rich perfume,
> And to your cheeks a fairer bloom;

With lustre teach your eyes to glow,
And health and cheerfulness bestow.'

Exercise not only preserves health and prevents disease, but aids greatly in relieving diseases even of the most obstinate character. Without exercise, medicine will fail to have the desired effect in a great measure.

SECTION III.
OF SLEEP.--

It is impossible for us to enjoy good health, unless blessed with sound and refreshing sleep, for without this tender nurse of weary nature, the whole frame is thrown into disorder, and the mind is much confused and weakened. When we are asleep, all the voluntary powers, such as seeing, hearing, feeling, &c., are in a state of suspension, or rest, while on the other hand, the involuntary powers, the circulation, digestion, &c., are increased, both in regularity and activity. A more uniform circulation is kept up throughout the system when asleep than when awake. I have often heard persons remark, with some degree of astonishment, that they would immediately begin to sweat on lying down and going to sleep in daytime; whereas, they might lie awake for hours on the same bed and not sweat. The cause is obvious, our several senses are at rest, and the circulation increased. The principal directions necessary to be given on this subject, are to take a proper portion of sleep at seasonable hours. The quantity of sleep necessary for each person every twenty-four hours, is hard to decide: it requires much more for some than for others. When a person rises in the morning, and does not feel refreshed, he may rest assured that he has slept too much or not enough. The best rule is to ascertain how much sleep you really need, and when you have obtained that quantum, rise from your bed immediately, and not lie dosing, and try to force yourself into sleep contrary to nature, for too much sleep and too little exercise, produce languor and debility: the nerves become relaxed, the flesh flabby and soft. Feather beds are unhealthy, especially in warm weather. A straw bed or matrass is much better for the health than feathers. A person wishing to enjoy good health should never retire to bed immediately after eating a hearty meal.

SECTION IV.
CLOTHING.--

Clothing should be suited to the age, constitution, and seasons. It should not be too warm in summer nor too cold in winter. All kinds of clothing should be made loose and easy, so as not to bind or cramp any part of the body: every attempt to give a good form by clothing is not only foolish in itself, but absolutely pernicious to health. Tight lacing not only obstructs the general circulation of the fluids, but oppresses the motion of the heart and lungs, and retards the wheels of life in the performance of their vital functions. The effects of tight lacing are bad health, coughs, indigestion, pleurisy, liver complaints, consumptions, &c.

Young persons need not be so warmly clothed as those who have passed the meridian of life. The weakly and those bordering on old age, should wear flannel. Wet and damp clothes should be particularly avoided: no fresh clothing should be put on without airing by the fire, no odds how long since it was washed. Many young persons injure their health by putting on damp clothes, lying on damp sheets, &c. Here much rests with their mothers; for such carelessness seldom fails to destroy the health, and often seats some incurable disease on the system.--These remarks are particularly applicable to young ladies who so often, when in a hurry, dressing for balls, churches, &c., risk their health and even lives, by putting on damp dressing, stockings, &c.

SECTION V.
OF FOOD AND DRINK.--

We cannot live without food and drink, and some attention to the quality of both food and drink, is essential to health. It would, however, be impossible to specify in this short work, the effects of every kind of diet, or to designate the quantity or kind of food, which will be most beneficial to the different constitutions.
"Diet may not only change the constitution, but it has been known to cure diseases, and it has this advantage over medicine, it is not disagreeable to take." Different constitutions require different quantities and qualities of food. The best directions that can here be given, are to be moderate as to quantity, and let the food be plain and simple, use only such diet as agrees with the stomach.--Eating of a single dish at a meal, is more healthy than indulging in a great variety. A diet composed of a proper mixture of vegetable and animal substances, will probably be found most nutricious and salubrious. Rich sauces, high seasoned provisions, where a variety of ingredients are intermingled, overload the stomach, and tend to produce dyspepsy. The flesh of young animals is more nutricious, and easier to digest than that of old ones. Persons whose constitutions are weak, ought to avoid eating food that is tough and indigestible. All rational persons who have arrived at mature age, are sufficiently acquainted with themselves, to know by a little attention what kinds and qualities of diets best agree with them: they should use such diets, and at such times as best agree with them; and if heads of families, they should pay some attention to what kinds of diets best agree with its different branches.

The best rules for eating, are to have your meals regularly, never fast too long or eat heavy suppers. Long fasts produce cholic, sick head-ache, costiveness, &c.--Break fast and dinner should be something substantial, supper should be light, and we should never lie down immediaely after eating.

As to drinks, pure water is of the utmost importance to health. Many persons think that it is pernicious to health to drink water before breakfast, but this is certainly a mistake. A reasonable portion of water taken before breakfast, prepares the stomach for food and facilitates digestion. Water however should never be drank in large quantities when over heated, as it is apt to produce disease, and sometimes immediate death. Coffee, tea, chocolate and milk, are all wholesome for such persons as they agree with, but must be decided by experience, as every person is best calculated to judge for himself.

SECTION VI.
OF CLEANLINESS.--

Cleanliness is too great a preservative of health, to be overlooked in a treatise on the art of preserving it. It clears the skin of impurities, and promotes perspiration; it will even, in many instances, cure cutaneous diseases: it prevents the communication of infection. In towns it should be the object of public attention, as many diseases owe their origin, as well as virulency, to the neglect of it.

Cleanliness, though not a virtue in itself, approaches that character, and should be observed with the greatest scrupulosity, and appreciated almost as a virtue. It is necessary to decency--it affords personal comfort, and is one means of rendering us acceptable to society. It is an evidence of gentility, regarded as necessary by the higher ranks of society, and is an ornament to every class; and without it neither health nor respectability can be long maintained. It is praiseworthy among those who enjoy good health, and still more important to those who are afflicted.

Cleanliness of the body is to be effected by changing the dress at proper periods, and by washing its surface. Frequent bathing braces the nerves and vivifies the spirits.-- Bathing is a powerful preserver and restorer of health; it softens and cleanses the skin, opens the pores, promotes perspiration, and invigorates the whole system.

SECTION VII.
OF THE PASSIONS.--

Man is a complicated machine, his soul and body mutually affecting each other. Much has been and might still be written on this subject: but as I do not intend entering into a general or scientific dissertation on the passions, I will confine my remarks to their influence on the physical system. The influence of the passions on the human system have long been observed, and sometimes remarkable cures have been effected by operating only on the mind. The restoration of tranquility and the diffusion of contentment and serenity, is often necessary, in order to give medicines a fair opportunity of having their accustomed efficacy. The subordinate indulgence of passion, frequently induces disease of a stubborn character, by destroying the power of digestion, enfeebling the circulation, affecting the brain and nervous system, &c. &c. But how mind and matter reciprocally act on each other, is a mystery which I leave to be developed by the researches of the profound philosopher.--When passions run counter to reason and religion, they produce the most frightful catastrophes. "When passion reigns reason is dethroned."

Young persons should early be taught to control their passions, as "the early management and control of the passions by a proper education, is the best guard against their mischievous effects at any period of life. When the habit is once established, their control then becomes comparatively easy; but when the curb of piety, reason or habit is not put on them, the ordinary excitements of unexpected circumstances, spurs them into a gallop."

OF ANGER.--

Anger is a sudden emotion of displeasure, excited by some supposed or real injury, offered either to our persons, characters or rights. Although anger is one of the most powerful and dangerous passions, both to ourselves and to the object of our wrath, yet we have as much or more power in governing it, than any other of the passions, to a certain extent. The intensity of this passion does not depend entirely upon the magnitude of the insult received, but also upon the pride, or rather vanity, of the individual who receives it. When an individual who has an exalted or overrated opinion of his own dignity and importance receives an insult, his vanity, like a magnifying glass, enlarges it into the most aggravated injury, and consequently, his ready resentment will equal the supposed magnitude of the offence. Persons addicted to violent and unrestrained fits of anger, are too often induced by the irritation of the moment, to perpetrate acts of the most alarming and outrageous character. Such deeds of rashness lead to the prison, and even the gallows.

Anger is a disease of the mind, a short-lived insanity, producing the rashest, maddest deeds of folly. This is the passion which has raised up nation against nation, which has destroyed millions of the human race, and desolated whole countries. It is even sometimes seen to deform the maiden cheek with a frown. It disqualifies its subjects for all kinds of business, or social intercourse with his fellow beings, and renders him miserable to himself and his associates. The storms of this passion have in some instances been so violent, as to produce immedtate death. Every passion grows by indulgence, and anger when unrestrained, is apt to degenerate into cruelty; and as self government and habit are the best preventatives of this dreadful and frightful monster, how early and cautious should examples of mildness and good humor be set before children by their parents. They should be taught to control this passion above all other things, for you may plainly observe the pernicious effects which anger produces on a child when indulged in it, as well as on a person of mature age.

"The exercise of patience is not only a duty, the performance of which prevents all the deleterious effects of anger, but it is an infallible mark of a great and dignified soul." Due attention to the formation of our habits will readily bring this passion under the salutary restraints of prudence and reason; but if suffered to rage without restraint, and to be blown into a flame on every occasion, it soon becomes ungovernable.

OF HATRED.--

This detestable passion is the voluntary fruit of a depraved soul: it is a voluntary and deep-rooted dislike, that seems to have its seat in the angry passions of the heart. Hatred is not in general, in consequence of provocation:--the object is not hated because it is odious, but because it interferes with inclination, &c. Hatred is a degrading passion; it is not contented with merely wishing evil to the objects of its fiendish malignity, but derives its only pleasure from their misery and destruction. Hatred has very appropriately been termed the "Bane of peace--the ulcer of the soul."

> "When hatred is in a bosom nursed,
> Peace cannot reside in a dwelling so accursed."

This detestable passion when permitted to occupy a place in the human breast, will soon make room for its sister passions--*Envy.* Slander, their offspring, will soon follow. Slander, whose mouth is ever full of lies, is truly said to be the "foulest whelp of sin." Enmity, ill-will, ranchor, malice and spite, are modifications of this base passion--Anger. They seek the misery, and are delighted in the misfortunes and destruction of their objects. Aversion, detestation and the like, when kept in proper bounds, are allowable emotions of the soul; they are not personal feelings, directed against the object independent and regardless of its qualities, but they are emotions produced by actions or things:--thus the virtuous and honorable, detest the base, the treacherous, &c. Hatred, operating as it does upon the mind, cannot fail to injure the body. It destroys the happiness, and consequently impairs the health.

ENVY.--

Envy, like hatred, is a low, degrading, and detestable passion. It is ever blind to the virtues and accomplishments of others, but quick-sighted in detecting imperfections that none else can see. Envy, like anger. is the bane of peace, the ulcer of the soul. Solomon says "Envy is the rottenness of the bones." It is a sensation of uneasiness, accompanied with malignity, excited by the superior accomplishments or advantages of a rival. It has its seat or root in an overrated self-love and thirst for praise, desiring to be esteemed superior to others, without efforts to merit such esteem. It never seeks to excel a rival by the practice of virtues superior to his, but labors to degrade him to its own level. The means employed to accomplish this vile purpose, is Slander, and thus the three enemies to all the fair forms of truth, honor, peace and happiness, unite their fiendish powers to destroy both soul and body.

> "Envy commands a secret band.
> With sword and poison in his hand;
> Around his haggard eye balls roll,
> A thousand fiends possess his soul.
> The hellish unsuspected sprite,
> With fatal aim attacks by night,
> His troops advance with silent tread,
> And stab the hero in his bed.

Or shoot the wing'd malignant lie,
And female honors pine or die."

AVARICE.--

I copy the following able piece on this subject from the writings of A. H. Mathes:

Avarice is a sordid passion. It is a craving anxiety after property; a rapacity in getting, and a tenacity in holding it. It is a grovelling passion, that seeks for happiness beneath the skies, and expects to realize, by hoarding up perishing dust, permanent enjoyments. When this sordid passion fakes possession of the heart, farewell all sentiments of honor--all correct notions of honesty, the only rule of right, and measure of wrong with the miser in his own interest: no other argument can reach his selfish soul. Farewell to all natural affections, and all the objects of gratitude; it wrests the last drop of humanity from the bosom, and strips it of the last feeling of compassion. The shrill cry of justice, or the deep groans of want, are notes equally beyond his compass. He can behold misfortune's most afflicted sons, driven by adversity's fiercest gale, wrecked on the ocean of poverty, with scarce a broken piece of the wreck, to buoy their heads above the waves of utter want, without one pensive reflection.--Without a sigh he can strip nakedness of its rags, and rob poverty of its crusts, or enter the forlorn cabin of the widow, and exact the uttermost farthing, leaving her fatherless babes breadless.

Avarice unties the bonds of society, and robs the miser of one of the greatest blessings in it--the mutual communication of kind offices. It dries up the fountain of humanity, obliterates every sentiment of generosity, and freezes up every stream of sympathy. As soon may you expect to pluck the blooming rose under the frozen pole, as to find the warmth of affection in the miser's frigid heart. From a region so barren of virtue, men no more expect to reap the fruits of charity, than they expect to gather grapes from the thistle, or figs from the bramble. This base passion robs the man of content; for although nature is content with few things, avarice is not content with all things; it tortures the soul and wastes the body with craving anxiety. His thievish fancy hears in every sound the approach of the robber. Of all the sons of folly, who barter time for eternity, life for death, heaven for hell, none do it on easier terms than the wretched worshiper at mammon's shrine, "who to the clink of mammon's box gives most greedy and rapacious ear;" is the only music that can charm him. Avarice renders a man poor in the midst of wealth, his niggard soul can scarce allow a scanty supply of food and raiment to his body, and for fear of future penury, reduces himself to present and utter want.

"And oh! what man's condition can be worse
Than his, whom plenty starves, and blessings curse?
The beggars but a common fate deplore.

> The rich man is emphatically poor,
> If cares and troubles, envy, grief and fear,
> Be the bitter fruits that fair riches bear;
> If utter poverty grows out of store,
> The old plain way is best--let me be poor."

Avarice is accompanied with extreme eagerness to make money, with distressing fears about keeping it, and with inconsolable grief for fear of losing it; besides heart-ache, envies, jealousies, sleepless nights, wearisome days, and numberless other ills which it inflicts on its slaves, ruining their health, and dragging them to the grave with some wasting malady, or hurrying them there by rash, horrible suicide."

The miser on being disappointed in an advantageous trade which he had thought almost confirmed, and fancied himself in possession of his new treasure; in losing the best of the market for his produce; or in having his hordes robbed of their idolized and shining dust, has, in many instances, been so smitten with grief, as to produce insanity, or rendered life so burdensome as to induce him to commit suicide.

FEAR.--

Fear was given to man as a sentinel of self-preservation. It induces us to take measures to avert, if possible, the apprehended ill, and secure personal safety.-- Apprehension, dread, &c., are modifications of the same passion. We apprehend what is possible, fear what is probable, and dread what is certain.

Fear has a salutary influence in society, amongst those who are now governed by the principles of virtue. The fear of reproach, punishment, &c., often rertrains the hand of violence, injustice and oppression.

Fear like every other passion, is liable to excess, and when thus indulged, instead of warding off anticipated evils, it often brings on the very calamities which are so much dreaded, and becomes hurtful to both body and mind. Fear indulged to excess, robs its possessor of resolution, reflection, and judgment, and degenerates into cowardice, which is a base passion, and beneath the dignity of man. No passion has a greater tendency to produce and aggravate disease than fear, when improperly indulged. It impedes the circulation, disorders the stomach and bowels, enfeebles vital action, and has a direct and instantaneous tendency to produce spasms on the whole system; and instances are not wanting, in which a sudden and excessive fright, has produced immediate death.

The practice of frightening children and grown persons, is often productive of the most deleterious consequences. Children are often fearful in the dark. This should be overcome by persuasion and argument rather than force. By proper treatment on the part of the parents or nurse, such unfounded fears will soon vanish; on the contrary, if they are encouraged by dismal stories of witches, ghosts, raw-head and bloody-bones and the like, they will grow and become so deeply rooted, that to shake them off in mature years, will be almost a matter of impossibility.

HOPE.--

Hope is an enlivening passion, it is a pleasurable emotion of the mind, excited by the anticipation of some desirable object considered attainable. It matters not, in what vocation we embark, it is our anchor to the last breath. We are supported by it in every difficulty.--It is hope with its offspring, *fortitude,* that enables us to bear all the toils, tumults, pains and vexations, which we have to encounter while passing through this "world of thorns." "It is the first friend that offers solace to the sons of affliction: it is the last to forsake them." Take from us hope, and life itself would be a burthen!

Hope is productive of the most salutary effects, both on our bodies and minds, differing in this respect from all our other passions. When engaged in the pursuits of life, and enjoying ordinary health, it is attended with many favorable effects without possessing any physical disadvantages, and what a powerful effect it has when laboring under pain and diseases of the body! It raises the spirits: it increases the action and power of the heart; gives vigor to the nervous system; moderates the pulse; causes breathing to be more full and free, quickens all the secretions and gives tone and strength to the whole system. The christian's hope extends beyond this *vale* of *tears* and enables him, in the last struggle of death, to cry out: "Oh! grave where is thy victory, Oh! death where is thy sting."

JOY.--

Joy is a high degree of pleasure, excited by the attainment or possession of some desired good--the reception of good news, &c. &c. Delight, gladness, mirth, cheerfulness and the like, are different modifications of this passion. Joy is pleasure at high tide. When indulged in moderation it has a salutary effect on both the body and mind; but if it should be excessive or very sudden, it frequently does serious and lasting injury to persons in good health; and instances have occurred, in which it produced immediate death. Persons of an ardent, lively temperament, and of delicate nervous sensibility, are mort liable to suffer serious or fatal consequences from sudden transports of this passion. Precautionary means should be used to prevent such sudden transports of excessive joy, by preparing the mind gradually to meet its emotions, and by this means its dangerous effects will be obviated.

SECTION XI.

GRIEF.

This depressing emotion of the mind, is produced by the suffering of some calamity, or by sustaining the loss of something that contributed to our happiness. The intensity of the passion is generally in the proportion to the estimate we place on the object lost. Indulged grief often becomes settled melancholy--its victim sinks into despair and fatal insanity. Sorrow, grief, melancholy, despair, &c., seem to be different modifications of the same passion. Whether grief proceeds from real or imaginary causes, the destructive influence is the same on the healthy action of the system. It destroys the digestive powers--oppresses the lungs and weakens the nerves--it produces sleepless nights, headaches, weak eyes, costiveness, palpitations of the heart and not infrequently insanity and death. How frequently do we see our fellow mortal weighed down with this depressing passion, their pale and furrowed cheeks tell us they are sick! ah! and of what? of every thing and nothing!! They apply for medical aid--take medicine without weight or measure; but all in vain.--They are still sick--the contents of an apothecary shop will not give relief. The mind is the part diseased; and until the cause is removed it will bid defiance to the powers of medicine. In such case much rests with the sufferer. The cause should be removed if possible; if this cannot be done, we should remember that this is "a world of sorrow." And why destroy both health and happiness, by grieving about a thing we cannot help? We should exercise firmness and resolution, and reconcile as far as possible, the circumstances and condition to our wounded and oppressive feelings. "We should seek in piety those unwithering consolations which can sustain the mind under the severest strokes of adversity. From this source issue streams of living pleasure that cannot be dried up by the occurrence of disastrous events."

LOVE.--

As this passion is not productive of any bad effects on the health, when of a proper kind and properly controlled, there need be but little said on the subject.--Love is one of the master passions of the soul, when kindled into ardor. It exercises an uncontrollable dominion over all the powers of man. Pure and reciprocal love is one of man's most endearing delights--it is not wrecked by the storms of adversity nor starved out by poverty.--We are commanded to exercise this passion in Holy writ, which is a sufficient proof of its excellency. Thus we are commanded to love our parents, our companions and children, and even our enemies; and above all--our God and heavenly things. When this passion is confined within its proper limits, with due regard to its objects, it has a salutary influence on the mind of every rational being. The influence of propitious love is salutary upon the physical system, as it promotes all the secretions--invigorates the action of the heart--imparts vivacity to the spirits and brightens the countenance with cheerfulness.

Some writers when speaking on this subject, digress from the true intent of the matter and fall into a discussion of most of the other passions. Under the title of "Disappointed Love," they discuss at great length the beautiful effects of grief, jealousy, rage, reveuge, despair, &c.--Love, crossed or disappointed by the inconstancy or falseness of the beloved object, not unfrequently begets one or more of the above passions, and produce some of the wildest storms of passion that infest the seas of life, wrecking both happiness and health,

"Earth has no rage like love to hatred turned; And hell no fury like a lover scorned."

REMARKS PRELIMINARY TO THE MEDICAL PORTION OF THIS WORK.

That a knowledge of diseases is necessary to their cure, will be readily acknowledged; but a great difference of opinion prevails amoug mankind as to how this knowledge should be obtained. Some say it should be the result of personal experience; while others contend that education and theory alone, is all that is necessary to make a skillful physician. The union of observation, with the deductions of theory, will probably be found to lead to the safest conclusions. A medical education should be united with experience. Every disease is to be known by its peculiar symptoms, and the sagacity of the physician will be exercised in discriminating between different diseases by their different symptoms.

Regard should always be paid to the constitution, manner of life, age, sex, temper, &c. Some constitutions are peculiar and require a peculiar treatment. It would be injudicious to treat the tender, delicate and sickly in the same manner as the hardy and robust.

Females have diseases peculiar to themselves--their system is more tender and irritable and demands greater caution.

The following enquiries should be made previous to administering medicine to a sick person.

When were you taken? How were you taken? To what disease are you most liable? Is the disease constitutional or accidental? Are you temperate in eating and drinking? What has been your general health? What were your feelings for several days previously to being taken, &c.? If the patient be a female, you should also ascertain whether she has been regular in her monthly periods? Whether there is any suppression of urine, &c.

A physican, on entering the room of a sick person, should be easy and affable in his manners, and wait patiently the subsiding of any strong excitement his presence may have created.

THE PULSE.--By the *pulse* is meant the beating or throbbing of an artery, which is occasioned by the motion of the heart in propelling the blood through them. This motion of the heart and arteries is spoken of at greater length in the anatomical part of this work. The physician derives great information as to the condition of his patient, from knowing how the blood circulates. This is ascertained by feeling the pulse. The pulse in different persons varies, it beats quicker in the sanguine than in the melancholly--in the young and vigorous, than in the old and declining--children have quicker pulse than adults. The usual standard of a healthy indication by the pulse in grown persons is from 66 to 80 strokes in a minute.

Good health is indicated by a strong, firm, regular pulse.

1. When the pulse resists the pressure of the finger, feels full, and swells boldly under the pressure, it is called a full strong tense pulse--if slow and irregular, it is called a weak, fluttering and irregular pulse.

2. When the pulse feels like a string drawn tight, and gives considerable resistance to the presure of the finger, it is termed a hard, corded pulse.

3. The soft and intermitting pulses give their own meaning by name; they are very easily distinguished from each other, as in cases of great weakness of the system and a languid circulation, or on the approach of death.

4. An intermitting pulse is sometimes produced by oppression of the stomach and bowels, it also arises in some instances from an agitation of the mind. A *vibrating pulse,* with quick, weak pulsations, acting under the fingers like a thread, quick but very weak and irregular, indicates a highly dangerous state of the system. This pulse is generally accompanied with deep sighs, difficult breathing, and a dead, heavy languor of the eyes.

The above directions will enable any person to distinguish the different states of the pulse; and enable him so far as the pulse can give any indications, to judge of the nature and stage of the disease.

PART TWO.
CONTAINING A DESCRIPTION OF THE SYMPTOMS OF DISEASE, AND THE METHOD OF TREATMENT.

I will not trouble the reader here with a long treatise on the doctrines, or what is called the *pathology* of diseases, which would prove both tedious and tiresome, without imparting the least benefit to those for whom this work is especially intended. But in giving the symptoms of disease, or the various aspects under which it makes its appearance, I will endeavor to do it in both a concise and comprehensive manner, confining myself, principally, to

those diseases which are most common in our country, and peculiar to our climate. It is very necessary that the head of every family should be instructed, to some extent, in the method of curing their own maladies; simple remedies, and such as are at hand in most families, will, if taken in due time, often throw off diseases which might have baffled the skill of the most experienced physicians if it had been let to run on without remedy for a length of time. A full conviction of this fact will induce me to simplify the HEALING ART, so that any family, possessing an ordinary share of common sense, may become their own physician in most cases of disease, without the hazard of increasing the hold of disease or weakning the power of life. The Indian system of practice may appear simple to many persons who are not acquainted with their success in treating diseases, but I flatter myself that a fair trial of their method of treating diseases as is herein laid down, will almost invariably be crowned with success, and many painful and truly distressing complaints which have hitherto been considered by the whites, as incurable, will be found to yield speedily to simple remedies.

Believing that colds are directly or indirectly the cause of most diseases by checking perspiration, obstructing the general or natural circulation of all the fluids, and there-by producing a marked action, or in other words, a diseased condition of the whole system--I will first begin with colds.

CATARRH OR COLD--(Oo-hur-tlah.)

Colds are so common in every country, and their modes of treatment so generally known, that the reader will doubtless conclude that little or nothing need be written on a subject which is already so familiar. But when we reflect that it is often the forerunner and not unfrequently the foundation of other diseases which are difficult to remove, and in many instances highly dangerous, and even fatal, in despite of medical aid--the subject does not appear so trivial as on first thought; but is one which certainly demands the serious attention of all those who wish to enjoy a reasonable portion of health.

Persons of delicate constitutions are most liable to take cold--and from the great carelessness of such persons in neglecting to avoid exposure--and to remove cold in its

earliest stages, originates most of the consumptions in this, as well as other countries. It is often the foundation of a host of other diseases, such as pleurisy, liver complaints, fevers, asthmas, &c. I therefore feel it my duty to impress it on the mind of the reader, that cold, however simple it may at first appear, should be taken in its earliest stage, at which time it is easily thrown off and by very simple means.

SYMPTOMS.--A dull heaviness in the head, frequent sneezing, a discharge of watery mucous from the nose or eyes, or both, a stoppage in the nose and head--it is frequently attended with chillness, succeeded by flushes of heat, a very disagreeable fullness is felt about the eyes. Cold is often attended with soreness of the throat, cough and pain in the chest. Here I repeat that most of the consumptions of this country are occasioned by neglected colds, brought on by exposure to night air, by changing warm clothing for thin, by sudden check of perspiration, by damp feet, &c.

TREATMENT.--Cold in its first stage may be thrown off very easily, and by very simple means, such as a free use of sage, mint, ground ivy, balm, pennyroyal; pepper or ginger teas, or any sweating tea that the patient may prefer, to which may be added a portion of the diaphoretic drops. If the violence of the attack requires it, bathe the feet in warm water fifteen or twenty minutes, then wipe them dry and draw on warm stockings. If the head should be much stopped up with cold, sweat it by covering it over with flannel or other covering, and place a hot rock on the hearth, then sprinkle water and vinegar on the rock, at the same time holding the head over it. After steaming the head in the above manner, care must be taken to avoid exposure to a free current of cold or damp air, which would check the perspiration suddenly, and in all probability do much more harm than the steaming had done good. If the symptoms are inflammatory, give cooling purges, such as cream of tartar, salts, castor oil, rheubarb, or any cooling cathartic. If the throat is sore, apply the red pepper poultice, or a poultice of onions or garlic, either of these poultices will give relief to the breast if applied to that part, in case of oppression from cold. If the patient is troubled with a cough, look under that head for a remedy; by turning to the index you will be referred to numerous valuable articles for coughs, some of which can be easily procured in all cases, I presume, with but very little trouble or expense. The onion, garlic or pepper poultice, applied to the feet will also aid in produing a free perspiration. The following remedy, says Dr. Gunn, "has frequently afforded relief in cases where colds had nearly settled down into confirmed *consumptions*--take one tea spoonful of flaxseed, half an ounce of liquorice, and a quarter of a pound of raisons, put them into two quarts of rain water, and simmer the whole over a slow fire until you reduce the quantity to one quart; then prepare some candy made from brown sugar, and dissolve it in the quart of liquor. A half pint of this syrup is to be taken every night on going to bed, mixed with a little good vinegar to give it a slightly acid taste. This will certainly relieve a cold in a few days." The vinegar stew is also very good for colds, and should be prepared in the following manner: If the vinegar be very strong, add a little water, then put it on the fire until it becomes hot, then add a little butter and sweeten it well with honey This stew or syrup is good to relieve soreness in the breast, it is also good to check the cough arising from cold. A tea spoonful of paragoric or half that quantity of Bateman's drops may be added to the tea, which is to be drank for cold to great advantage.

INFLUENZA OR MALIGNANT SORE THROAT.
(Oh-ch-tlah-tsu-ni-sik-wah-his-lee.

This dangerous disease is sometimes called putrid or ulcerous sore throat. The symptoms are, soreness of the throat, attended with fever. The swallowing becomes more and more difficult, the skin burning and disagreeably hot without the least moisture, the pulse very quick and irregular, it is also attended with nausea and sometimes vomiting, restlessness, great debility, the face becomes flushed, the eyes inflamed, and the neck stiff, the mouth and throat assumes a fiery red color, and the palate and glands of the throat become much swelled as the disease advances, the whole internal surface of the mouth and throat will become interspersed with brown or ash colored spots, which soon become so many ulcers discharging an acid matter; a similar matter runs from the nose, and escapes at the mouth, this matter soon affects the lips and neighboring parts, and in some instances the brown spots extend over the whole body, the tongue becomes covered with a thick brown fur and the breath very offensive; there is generally a purging, and in many cases, a frequent discharge of excoriating matter or fluid from the fundament. If the disease is not checked, the ulceration corrodes deeper and deeper extending down the alimentary canal, and if still suffered to proceed, they become gangrenous; a severe purging ensues, and death closes the painful scene. The following symptoms are unfavorable and denote a fatal termination; the feet and hands become cold, the eruptions suddenly disappear, or become of a dark livid color, the inside of the mouth and throat assume a dark hue, purging a black matter of a very offensive smell, the pulse becoming small, quick and fluttering, hurried breathing with frequent sighing; and a cold clamy sweat. On the contrary, the symptoms are favorable when the fever in some degree abates and the skin becomes gradually soft and moist, the breathing becomes more free and natural, the eyes assume a natural and lively appearance, the eruptions on the skin become of a reddish color over the whole body, and the parts which separate from the ulcers fall off easily, and leave the sores of a clean and reddish color, the tongue gradually becomes clean and clear of the dark fur with which it is covered. These are favorable symptoms and denotes the recovery of the patient.

Putrid sore throat, is an infectious or catching disease; and hence it sometimes prevails as an epidemic, and generally makes its appearance in the fall or early part of the winter seasons, especially when preceded by a dry, hot summer. Children and persons of delicate constitutions are most liable to be the victims of this dangerous disease. Neglect of cleanliness, eating damaged provisions, breathing impure air, or whatever tends to produce putrid fevers, will predispose to an attack of this complaint. When relief is not had, this disease generally terminates fatally between the fourth and seventh day.

TREATMENT.--This disease generally makes its appearance at the close of sultry seasons, when the system is much weakened by protracted exposure to intense heat and when people have been for some time exposed to breathing the putrid atmosphere arising from stagnant waters and decaying vegetables. This fact will at once show the impropriety of administering sever purges or drawing blood. The stomach must be cleansed by an emetic of gulver and Indian physic, and the bowels relieved of their putrid contents by injections, of thin gruel or soap-suds, to which may be added hogs lard and a little gulver syrup; no cathartic stronger than castor oil or rheubarb should be taken into the stomach. Well prepared charcoal, taken twice or three times a day, will be of great benefit. The mouth and throat must be washed and gargled with a preparation made as follows: Take of cayenne pepper in powder, two table spoonfuls, a small quantity of catnip and half a spoonful of common salt: pour on them one pint of boiling water, let them stand a half hour and strain off the liquor and add to it a half a pint of good vinegar--the patient should also swallow a table spoonful of this preparation every fifteen minutes. If the patient should become very weak, bathe him well in a strong decoction of red-oak bark, in which may be put one-fourth whiskey. If the weakness be very considerable, give wine or toddy made with spirits and sweetened with sugar to strengthen and support the system. For an external application to the

throat, use a poultice made by thickening rye-meal or wheat-brand in red pepper tea.--
After the stomach is cleansed, give Virginia snake root tea, (commonly called black snake-root,) or seneka snake-root tea freely. The bowels must be kept regular through the whole course by the use of injections. If the first emetic should fail to subdue the disease it should be repeated in moderation on the day following. By properly attending to the emetic, the acid matter may be thrown off, which would otherwise produce injury by descending into the bowels. The strength of the patient must be supported by a generous, nourishing and easily digested diet, comprising but little if any animal food.

PLEURISY.--Oh-ne-squah-ga-ni-tsu-na-his-na.

SYMPTOMS.--An accute pain in the side, extending to the back, breast and shoulder, when the breath is drawn:--The pain is much increased by a short dry cough which generally attends it. Great difficulty is experienced in lying on the affected side. It is also attended with chills and fevers, great thirst and restlessness as in the inflamatory fever. The tongue is covered with a thick whitish fur. The urine is high colored, the face flushed, and the skin dry and hot; sometimes the cough increases, and a tough phlegm is spit up. The blood when drawn from the arm and admitted to cool in the vessel, is covered with a scum or coat of a buffly color, which always denotes inflamation.

The causes which predispose to an attack of this disease, are cold, lying on damp beds, exposure to free currents of damp air, wearing wet or damp clothes, sudden changes from heat to cold, sudden check of perspiration, suppression of periodical evacuations, or by the repulsion of eruptions. It may also arise from intemperance, great exertion in singing, speaking or playing on wind instruments

TREATMENT.--It is an inflammatory disease, and therefore requires the immediate reduction of the inflammatory symptoms; for this purpose bleed freely according to the strength and constitution of the patient. If the pulse should remain full and hard after the first bleeding, and the pain be relieved for a short time and then return, you must bleed a second, third, and even the fourth time, where the inflammatory symptoms require it. After the first bleeding, apply cloths, wet with hot water to the pained part, as warm as can be borne, and bathe the feet in warm water. At the same time give a purge of epsom salts or gulver pills, and let the patient drink freely of a tea made of one-third of silk-weed root to two-thirds pleurisy root. If this tea should increase the fever to any considerable extent, it may be used in smaller quantities and the lancet again resorted to. For a description of the above roots, look under their different heads You will also see the mode of preparing the black or gulver pill under its proper head.-- After the inflammatory action is in a considerable degree overcome, seneka snake-root should be combined with the silk-weed root and pleurisy root. A full description of these roots may be seen under their proper heads. After the abatement of the fever, if the pulse should sink and the patient become very weak, you should stimulate him with warm toddy or wine, mixed with warm water and sugar. This must be done with the greatest caution, taking great care not to stimulate so as to produce a return of the fever. If the extremities should become cold, apply plasters of ground mustard-seed wet with vinegar to the wrists, ankles and feet. These plasters will aid greatly in raising the pulse, and is not so apt to produce a return of the inflammatory symptoms as a too free use of spirits. The bowels must be kept open through the whole course by cooling purges; such as salts, castor oil, cream of tartar, or gulver pills. The cathartics should be aided by mild and cooling injections, such as thin gruel, well strained new milk and water, &c.

For further information on this subject examine under the head of "Clystering Diets." The strictest abstinence from all kinds of animal food, must be observed in this disease. The diets and drinks must be such as will have a tendency to keep down fever, and such as the stomach will most easily digest. The drinks should consist of flaxseed tea, slippery-elm tea, toast-water, &c. They should be taken warm, a little gruel, panada, or milk and water with mush, may be taken for nourishment.

When recovering from this disease, great care must be taken to avoid sudden changes, dampness, cold, and very particularly avoid exposure to night air, excessive use of ardent spirits, violent exercise, &c. As the reverse of the above precautions generally produce dangerous relapses. Flannel, or some warm dress should be worn next the skin.

DROPSY--Tsa-no-tis-scoh.

Dropsy is a disease of the whole system, arising from debility or weakness. This opinion is sustained by many of the most distinguished physicians in the United States. Dr. Rush was of opinion, that dropsy was caused by a morbid action of the arteries, and an increased action of the exhalents; or in other words, by an inactive state of the arteries and an active condition of the vessels which throw off the sweat from the body. Dr. Shelton's opinion is the very reverse, he says: "Notwithstanding the great popularity of this opinion and the high regard I have for Dr. Rush, yet I cannot concur with him. I believe the cause to be an *increased action* of the arteries and a *decreased action* of the exhalents. For we generally find in a Dropsy a quick pulse, which certainly indicates an increased action of the arteries; from the great fullness and distention of the exhalents, we might reasonably suppose they were too much relaxed, or too inactive to throw out the fluids as fast as they were forced into them by the active motion of the arteries." I have given the opinions of the above writers for the reflection and entertainment of the reader. The opinion of Dr. Shelton, however, accords nearest with my own.

SYMPTOMS.--Dropsy may easily be distinguished from other diseases, by the collection of water in some part of the body, and by the feet and ancles swelling, the flesh will have lost its elasticity, or in other words, when pressed upon by the finger the mark or impression will remain for some time after the finger is removed, the place where the impression was made being much paler than any other part. Among physicians, it is called by different names, according with the different parts of the system in which the water is deposited. When the water is seated in the cavities of the head or brain, the disease is called by physicians, Hydrocephalus. When seated in the cavity of the *chest,* it is called Hydrothorax; when in that of the *belly,* Ascites; when seated in the *scrotum* or *bag* of the privates, it is called Hydrocele; and when the water collects in the cellular membrane, which is situated between the flesh and skin, it is called Hydrocele. These different locations of Dropsy are manifested by somewhat different symptoms.

Anasarea or Dropsy of the celular membrane, first gives symptoms of its approach by swelling of the feet and ankles; this swelling may be distinguished from other swellings in the manner above stated. The swelling extends by degrees to the thighs, trunk of the body, and finally to the head and face. The breathing becomes difficult, particularly when the patient lies down. A cough soon follows, and a watery mucous is spit up, the urine is high colored, and is voided in very small quantities, and when suffered to remain in the urinal or pot it deposits a reddish sediment; the bowels are costive, and the thirst great. These symptoms are succeeded by a dull torpor and slow fever.

Ascites, or *Abdominal Dropsy,* is generally preceded by a loss of appetite, sluggishness, dryness of the skin, thirst, oppression of the chest, cough, decrease of urine, a swelling of the abdomen takes place, which increases gradually, as the disease advances. As the water accumulates, the breathing becomes more difficult, the countenance pallid and bloated, the thirst immoderate; the urine scanty, high colored; and deposits a brick colored sediment.

Hydrothorax, or *Dropsy of the Chest,* generally comes on with a sense of uneasiness at the lower end of the breast bone, and difficulty of breathing, which is much increased
by exertion, or by lying down. It is attended with a cough. at first dry, but afterwards a thin mucous is spit up; as the disease advances, the thirst increases; the complexion becomes sallow; the feet and legs swell; the urine is voided in small quantities, high colored, and deposits a red sediment. The face and extremities become cold, the pulse feeble, and irregular; the sleep is much disturbed, frequent palpitations of the heart; a numbness extends from the heart towards one, and sometimes, both shoulders; the difficulty of breathing continues to increase until death ends the patient's sufferings. Hydrocephalus, or *Dropsy of the Brain,* is a disease common to children, and will be spoken of under the proper head.

TREATMENT.--Cleanse the bowels with anti-bilious pills, or some other purge. After the bowels are well cleansed, the patient should take the diuretic pill night and morning, three for a dose, or more if the constitution of the patient requires it; also drink bitters by putting a table spoonful of steeldust, and about four ounces of vervine root, into a half gallon of good spirits; of these bitters the patient should drink three or four times a day what the stomach will bear.

ANOTHER REMEDY.--After the bowels have been cleansed as above directed, let the patient drink freely of cold water off of the root of *Ah-squah-na-ta-quah.* This herb is fully described in *materia medica,* and is an infallible remedy for Dropsy; the root should be bruized before it is put into the water, about a half ounce of the root to a quart. the water may be renewed until the strength is all extracted. There are no disagreeable consequences whatever produced by the use of this root.

The Chalybeat pill, taken night and morning after the bowels have been cleansed, will effect a cure in most cases. A Dose in this case is one pill about the size of a summer grape.

Diets must in all cases of Dropsy be of the lightest and simplest kind. When the patient begins to recover from Dropsy, the appetite in most cases becomes voracious and almost insupportable, and if the patient is permitted to indulge it, to effect a cure will be found impossible. Water gruel, rye mush and butter milk or something of the kind is the safest nourishment I have ever tried in cases of this kind.

After the patient is freed from the water, extreme debility usually takes place. At this stage of the disease, the patient should continue whichever of the above treatments may have been adopted, and additionally use wine and a decoction of wild cherry-tree bark, or a decoction of Columbo root, or any other stimulant or tonic that may be most convenient.

Dr. Gunn believes Dropsy to be an inflammatory disease and recommends bleeding freely, but goes on to speak highly of the advantages that have been derived from herbs of our own country, in this as well as other diseases. I quote the following from this author: "The following cures, which I shall notice, in the words of an experienced and distinguished physician, give evidence of the correctness of some of my introductory remarks, among which are the following: The discoveries of each succeeding day convinces, that the Almighty has graciously furnished man with the means of curing his own diseases, and there is scarcely a day, month or year which does not exhibit to us the surprising cures made by roots, herbs and simples, found in our own vegetable kingdom, when all foreign articles have utterly failed. The truth is, that the wise and beneficent Creator of the Universe, has made nothing in vain; and the time will come, when the apparently most useless and noxious plants, will be found eminently useful in the cure of diseases, which have hitherto baffled the profound skill, and most powerful energies of genius."--The following are the words of the author just alluded to: "I am knowing to two extremely distressing cases of Dropsy entirely relieved by means of the bark of common Elder. One a woman advanced in age, in the last stages of this disease, who lost a brother some short time previous, by the same complaint. The other a young woman, who had been for eighteen months confined to her bed, during four of which she was unable to lie down, and who is now wholly free from Dropsy, and recovering strength in a most surprising and unexpected manner. This young lady used the elder-barked-wine, at the instance of one of the most distinguished physicians of Boston, who had previously tried every known prescription without success, and the use of the elder bark entirely cured her. A great many other cases, less aggravated, have been cured by this bark. I have used it myself with unusual success, and its immediate adoption by the afflicted, is truly important and deserving attention. The receipt is as follows: Take two handfuls of the green or inner bark of the white common elder, steep it in two quarts of Lisbon wine twenty-four hours--if this wine cannot be had, Teneriff or Maderia, will answer; take a gill every morning, fasting, or more if it can be borne on the stomach."

We have never tried the above preparation of elder bark in wine, but having witnessed similar effects produced by the free use of the tea and decoction of this bark, we are bound to place full confidence in the above statements, and earnestly recommend its use to those who may be afflicted with this truly distressing complaint.

Diets should consist of gruel, a little milk and mush, or something of a similar nature.

DYSPEPSY, OR INDIGESTION.
(Oh-ne-na-tse-tsunah-li-stoo-na.)

SYMPTOMS--Are flatulency, defective appetite, palpitations of the heart, painful distention of the stomach and bowels. The last named symptoms greatly increased by eating a hearty meal or drinking spirituous liquors. This disease also extends its pernicious influence to the mind, which often becomes desponding and irritable, and the poor sufferer exhibits a peculiar anxiety of countenance. The sleep becomes disturbed and the urine high colored.

CAUSES.--This disease originates in a great variety of causes. It arises in a great many instances, from a diseased state of the Liver as may be fully seen under that head. This lingering and painfully distressing malady is seldom to be met with among the Indians, owing, we suppose to the great simplicity of their diet, and the liberal exercise which they so generally take in the hunt, the chase, &c.; and the little use made by them of mercury in any form, or of strong minerals of any kind. This disease, on the contrary, appears to increase yearly among the whites. It seems to be a scourge upon the more refined portion of the human species, and one which refinement with all its charms, utterly fails to render agreeable, or in any respect desirable. It is to be found among all ranks and sexes; but when we meet with an individual who is afflicted with this torturing malady, and examine into his or her past life, the cause is generally obvious. An excessive use of spirituous liquors of any kind, of tobacco, mercurial preparations, and other poisonous mineral substances used for medicines--improper diet, whether in kind or quality, inactivity of body, intense study, uneasiness, anxiety or grief, are all calculated in their nature to produce this painful disease. Dr. Carter, when speaking of the stomach, says. "It may be considered the great labratory or chemical workshop of the living power--where chemical operations upon our food and drink, are regularly performed, without effort, toil or study."

Dr. Carter's statement respecting the stomach, shows at once the great necessity of regulating the food and drink according to the strength of this "chemical workshop."-- The usual practice of over loading the stomach with high-seasoned, indigestible food, and a too free use of ardent spirits, injures its tone, and renders it incapable of performing its functions in a healthy manner. If Indigestion arises from a diseased state of the liver, there will also be felt a dead, heavy pain in the right side, also in the shoulder, and back of the neck. The urine on being deposited in a urinal or pot, will have in the bottom when cool, a reddish colored seddiment. The complection will become of a tawny or yellowish hue. The feet and hands, when held in one position, for a short time will become asleep for want of a free circulation of the blood. Great uneasiness will be felt throughout the whole system, and it is sometimes attended with vomiting. When these last named symptoms occur, you must refer to the proper treatment of the disease of the liver, &c.

TREATMENT.--For common Dyspepsy, the patient must first sum up all the resolution which it is in his power to command, in order to regulate his diet with that rigidness. which is indispensably necessary, where a cure is to be sought for in this disease--the diet must be simple, such as gruel, a little rice prepared in clear water, and salted just sufficient to make it palatable, a cracker with a glass of spring water, or some similar food. It must be taken frequently and in small quantities, as fasting too long does great injury in Dyspepsy, which injury is much increased by the common practice of over-loading the stomach after long fasting. The use of animal food must be entirely abandoned if the sufferer wishes to obtain relief. To be

alternately loading the stomach with purges, animal food, and spirituous liquors, is only adding fuel to fire, and will ultimately end in the death of the patient if persisted in. The patient should first take a purge, anti-bilious pills will be most suitable. These should be taken on going to bed--the number for a dose refer to that head--if they should not operate by morning they should be aided by half a dose of the same pills or by castor oil. After the bowels have been cleansed in the above manner, take a portion of the anti-dyspeptic syrup or hepatic pills morning and night. For directions how to prepare either of the above named medicines look under their different heads. While using the anti-dyspeptic syrup or hepatic pill, you should also use a mixture or syrup made by taking a strong decoction of the inside bark of white hickory, one pint well strained, to which add an equal quantity of soot, a pint or more of honey, of this mixture take a tea spoonful morning and night. A free use of charcoal, taken in water or otherwise, will be found of great benefit. For directions for preparing charcoal refer to that head. During the above course, the patient should take moderate exercise in the open air, if the strength will allow, and be very careful to avoid any thing either in eating or drinking, that will produce aggravation of the symptoms. The bowels, if they become costive, (which however is not apt to be the case while using the anti-dyspeptic syrup or hepatic pills,) must be regulated by the use of mild and cooling clysters. When the stomach and bowels have been kept free from irritation for a length of time by the above treatment; when the sleep becomes tranquil: the spirits revived, and the tongue assumes a clear and healthy appearance, a little mutton or beef soup may be taken. or chicken well boiled and the soup thickened with a little flour. If this diet should produce an uneasiness in the stomach or bowels, the quantity taken should be diminished, and if it still aggravates the symptoms, its use must be entirely discontinued, and recourse must again be had to the former simple dish--gruels, &c. But if the stomach will bear light meals of the above soups, the quantity may be gradually increased. but it must be done with great caution.

I have known several persons relieved of this distressing complaint by the following simple remedy, after other remedies had been tried and had failed: Take of cob ashes, steel dust, and common salts, of each a table spoonful, mix them well together, and add a sufficient quantity of honey to wet it or stick it together. Take of this mixture what will make three common sized pills morning and night, and noon if the stomach will bear it. At the same time take charcoal in water, prepared as directed under that head, and regulate the diet as before directed. I have known this to cure two cases of dyspepsy after the prescriptions of a physician in high standing had been tried and failed.

FLUX OR DYSENTARY.
(Gee-guh-tsi-too-nuh-goo-skah.)

SYMPTOMS.--A constant desire to go to stool without being able to pass much of anything from the bowels, except a bloody kind of mucous. These desires to go to stool are usually accompanied with severe griping, and also with some fever; as this disease advances, the stools will consist of pure blood and matter mixed; and from severe straining to evacuate, part of the bowels will frequently protrude or come out, which soon becomes a source of great suffering, it is also attended in many instances with chillness, loss of strength, a quick pulse, great thirst, and an inclination to vomit.

CAUSES.--Dysentary or Flux is generally most prevalent in the latter part of Summer and in the Fall, though it frequently occurs in other seasons of the year. A long drought followed by cold rains is apt to predispose the system to an attack of this disease. It is also produced by sudden suppression or stoppage of perspiration, which determine the fluids to the intestines; by eating unripe fruits; unwholesome, putrid food; and by breathing noxious vapors. Some writers say it is a contagious or catching disease, while others say it is not; be this as it may, it often attacks whole neighborhoods or towns at the same time; but it looks reasonable that the same general causes which produce it in a town, neighborhood or section of country, render all, whose modes of life and systems are in similar conditions, subject to it. This disease is more common in warm climates than in cold ones, and in rainy seasons than in dry ones. When it attacks persons of feeble constitutions or those laboring under scurvy, consumption &c. it generally proves fatal. Grert debility, voilent fever, cold clammy sweats, hickups, dark colored spots on the skin, coldness of the extremities, and a feeble irregular pulse, are symptoms of a fatal termination. This disease should be taken in its earliest stages, at which time it is easy to be subdued by the use of the proper remedies, but if suffered to run on it is sometimes extremely difficult to overcome.

TREATTMENT.--Take a handful of each of the following barks, red-bud horn-beam, (commonly called iron wood,) red-elm, sweet-gum and black-gum; also, a handful of yellow root and iron-weed root, make a strong decoction of these articles, and let the patient drink of it freely, a purge of the anti-bilious pills should be taken to work off the acrid contents of the bowels. Another very valuable drink for this disease, may be made from the inside bark of swamp white-oak--take one pound of this bark, pound it well and put it into a half gallon of cold water. This is an excellent drink to cool and heal the bowels. If the belly be hard and sore to the touch, grease it well with any kind of oil or lard, or apply poultices of catnip to it. Injections of peach-tree gum or cherry-tree gum, made by dissolving the gum in water until it forms a mucilage to which may be added forty or fifty drops of laudanum for grown persons, and less for children, will aid in allaying the irritation of the bowels--the injections should be used cold. Castor oil combined with Bateman's drops, paragoric or laudanum may be used to advantage in this complaint--for a dose refer to the table of medicines; for a full description of all the above barks and roots refer to their different heads. The drinks during this complaint must be of the mildest kind, such as slippery--elm tea, flaxseed tea, &c., and diet of the lightest kind, such as light soups, jellies, new milk thickened with flour, all kinds of fruit must be avoided.

I have known many cases of this disorder, among children, cured by the free use of a tea of vervine root, which grows in such abundance about our yards. On recovering from an attack of this disease, great care should be taken to avoid exposure, for fear of a relapse, which is generally very easy brought on by exposure, violent exercise or improper food.

DIARRHOE OR LAX.--(Tsu-ne-squah-lah-tee)

This disease is characterized by frequent and copious discharges from the bowels, unattended with fever, and has not the appearance of a contagious or catching disease as is the case with flux. It generally prevails among persons of weekly constitutions; persons advanced in years and those who have lived intemperately. Many are subject to its attacks from the slightest cold or exposure, which at all affects the bowels; and others are naturally and constitutionally of this habit of body. The appearance of the stools in this disease are very different at different times, depending in a great measure on the food and the manner in which it agrees or disagrees with the stomach and bowels. This disease is very often produced by worms.

TREATMENT.--When this disease has been brought on by colds, or sudden stoppage of perspiration or sweat, use the warm bath and drink freely of some diaphoretic tea, to produce a determination to the surface, (or gentle moisture of the skin,) paragoric or Bateman's drops may be used with the tea--for a dose see table of medicines. Where this disease is constitutional it frequently continues through life, if not relieved by medicines. Such persons should be particular as to what kind of diet they eat, and strictly avoid everything that disagrees with their stomach or bowels; they should guard against damp feet, damp ground, &c., they should make daily use of bitters, composed of swamp white-oak inner bark, red dog-wood inner bark, sweet-gum and cinnamon bark digested in old French brandy; in violent attacks the decoction recommended for flux should be taken until the violence of the symptoms abates. Slippery-elm bark or the root of common comfrey forms an excellent drink in this complaint. Injections of the same are also good. Where this disease is lingering and is attended with great debility, a raw egg taken of a morning on a fasting stomach will be found of great benefit. It should be taken in fresh spring water. In many instances a tea of flaxseed, slippery-elm, comfrey or vervine will entirely relieve it in a short time. When worms are supposed to be the cause of this disease in which case the breath will have a very foetid or offensive smell, treat the complaint for worms--see under that head.

The Cherokee Physician, or Indian Guide to Health

The Cherokee Physician, or Indian Guide to Health

The Cherokee Physician, or Indian Guide to Health

The Cherokee Physician, or Indian Guide to Health

The Cherokee Physician, or Indian Guide to Health

HEMORRHOIDS OR PILES.
[Tsu-nah-tee-kah-stee-tsi-kah-nu-go-gah.]

This complaint is sometimes hereditary, that is, it runs in families, and all ages and sexes are liable to it. There are two kinds of Piles originating from the same causes, and are distinguished as the *bloody* and *blind* Piles. The Piles are small swelled tumors, usually situated on the edge of the fundament; where these tumors break and discharge blood, the disease is called bloody Piles; but when the tumors discharge no blood, they are called blind Piles. There is usually a sense of weight in the back and lower part of the belly, giddiness in the head, sickness of the stomach, flatulency in the bowels and generally fever. Severe pain is experienced on going to stool, and small tumors may be felt projecting beyond the verge of the fundament; when these tumors break and discharge blood, the sufferer experiences intervals of ease; but when they do not break great agony is experienced during every motion, and great inconvenience is experienced in sitting down on a hard seat. In some cases, the lower end of the gut protrudes, (which means to come down) the length of two or three inches every time the patient goes to stool, and looks very raw and tender; this last case mostly occurs in children of weakly habits.

CAUSES.--Piles may be occasioned by continued or habitual costiveness, by frequent drastic purges of aloes, by riding a great deal on horse-back in hot weather, by excessive drinking, exposure to cold, suppression of some accustomed evacuation, and by the pressure of the womb on the rectum, when in a state of pregnancy.

TREATMENT.--Cold water is certainly one of the best applications that can be made either for a preventative or cure for this complaint. I do not believe that any person will be afflicted much with either bloody or blind Piles, who will bathe the fundament daily in cold water, say twice a day. I have known several persons relieve themselves of this painful disorder by this simple application. But I will proceed to give other remedies for the benefit of those who may prefer a cure not quite so simple, and one that will require rather more trouble than the former. Let the patient drink freely of a strong tea of yellow root. For a description of this root, look under that head. For an ointment, take mullen leaves, pound them fine and stew or fry them in fresh butter until the strength is extracted, then strain it through a cloth; with this ointment annoint the rectum or gut when it protrudes or makes its appearance. Persons who are much afflicted with Piles of either kind, will derive much benefit from sitting on a stool or bench of green white-oak, a portion of each day; it should be made as warm as can be borne previous to each time of being used. Many persons are troubled with violent and sudden attacks of this disease, having at times but a very few minutes warning, until the pain is almost insupportable. In this case, the patient may obtain immediate relief by applying cloths wrung out of water or vinegar as hot as can be borne to the fundament; flannel cloths would be preferred, they should be changed every few minutes, keeping a fresh or warm one to the parts until relief is obtained. A salve made from the leaves, seeds or roots of the Jimson or Jamestown weed, and applied as an ointment, is a speedy and certain remedy. The mode of preparing it is as follows: Take the leaves, seeds or roots of this plant, bruise them well and stew them in fresh butter until the strength is extracted, then strain and cool for use, with this salve annoint the fundament frequently. A decoction of any part of this plant is also valuable when applied to the fundament by means of woolen cloths. I have known several children severely afflicted with this painful complaint, which was produced by extreme weakness; in this case I use tonic medicines, such as wild cherry-tree syrup or dog-wood or poplar bark syrup, and bathe the child once or twice a day in a strong decoction of dog-wood and red-oak bark. After each stool the fundament should be anointed with the Jamestown weed (Jimson) ointment or clean hogs lard. In all cases of Piles, the bowels should be kept open by the use of very mild cathartics. I prefer the use of equal quantities of cream of tartar and finely powdered sulphur, taken in sufficient quantity to keep the bowels gently open. All persons that are addicted to Piles, should live on light and cooling diets, take moderate exercise, and bathe the fundament frequently in cold water as above directed.

CHOLERA MORBUS, OR PUKING AND PURGING.
(Tah-to-ne-tse-luh-ne-gah-slee.)

This disease usually attacks with sickness at the stomach, pain, flatulence, and severe pain or griping in the bowels. These symptoms are soon succeeded by heat, thirst, quickness of breathing, with a quick fluttering pulse and violent puking and purging. When the extremities become cold, the perspiration or sweat cold and clammy, the pulse irregular and changing, accompanied with cramp and hickuping, the case may be considered very dangerous and will soon terminate in death if relief is not obtained. It is generally too late at this stage to apply for medical aid.

This disease may be produced by an excess of bile--by the food becoming rancid or acid on the stomach--by sudden check of perspiration, or by a sudden stoppage of the menstrual discharge. It is produced in some instances by breathing damp air; by being exposed to inclement weather; by getting the feet wet, &c.; but in most instances it is occasioned by eating such food as disagrees with the stomach and bowels. Many very different modes of treatment are on record among the whites for this distressing complaint--some recommends a puke, others a purge, blistering, &c.; and some have even recommended scalding the stomach, where death is so near as not to allow time to draw a blister with Spanish flies in the common way. "I have," says Dr. Foreman, "although an Indian, been a personal observer of their different modes of treatment, and the little success which generally attended it, I have never experienced any difficulty of consequence in arresting this disease, when called upon in any reasonable time, and that too with very simple means. Instead of punishing the stomach, which is already tortured with agitation, by giving an emetic, my first step is to tranquilize or quiet it."

TREATMENT.--First give a tea of the Cholera Robus root, which will soon stop the puking. This root or plant is called by the Indians or Natives, *Sah-ko-ne-ga-tre-kee,* "but I have," says Dr. Foreman, "always called it by the name of Cholera Morbus root, when speaking of it to the whites, as this name came nearest conveying a correct idea of its medical qualities. I have never seen the whites use it except when directed to do so by the natives, and if they have any other name for it I do not know it." A full description of this plant may be seen under its proper head. When the violence of the puking has measurably subsided from the use of the above named tea, it will be necessary to clense the stomach and bowels. For this purpose take a purge of anti-billious or gulver pills, or some active cathartic. For the mode of preparing these pills, refer to their different heads. If the extremities become cold, bathe the feet in warm water, and apply plasters of ground mustard seed to the feet, ankles and wrists. After the puking and purging has abated, if the patient should become very weak, stimulate him with weak toddy, give nourishing diet and such as the stomach will easily digest. The rapidity with which this disease proceeds, requires the remedies to be promptly applied, for the disease is, generally speaking, highly dangerous, and soon terminates fatally, unless relief is speedily obtained. In cold climates this disease is most prevalent in the latter part of summer and beginning of fall, when there are sudden transitions from heat to cold; but in warm climates it occurs at all seasons. Persons who are subject to this sudden and dangerous complaint, should be very cautious as to what kind of food they indulge in; and should be very particular in avoiding the causes which produce it--for, by indulging the appetite and by exposure to the causes which produce it, the disease may return with redoubled violence and danger.

SCROFULA OR KINGS-EVIL.--(Oo-niller-oo-tah-ner.)

SYMPTOMS.--Small tumors appear behind the ears; under the chin they also make their appearance, in some instances about the joints of the elbows, ankles, fingers and toes; rarely on other parts of the body. As these tumors grow larger, the skin which covers them, becomes of a purple or livid hue, with inflammatory symptoms; at length they break and become ulcers, from which is discharged a white matter some what resembling curdled milk. Young persons are most liable to become the victims of this disease. It is said by some writers, that "true scrofula never makes its appearance after the age of thirty, unless it has shown it self in some shape before." It is caused by a taint or constitutional weakness in parents; or from cold, strains, bruises, &c. Children of lax fibers, with smooth soft skin, fair hair and delicate complexion, are more liable to attacks of this complaint than those of a different character.

TREATMENT.--The existence of this complaint in any person, is a plain indication of a corrupt, morbid state of the fluids of the body. It must therefore be obvious, that the proper mode of treatment will be, first to correct and purify the fluids, this will prevent in a great degree the formation of other tumors, and aid external remedies in effecting a cure of those already formed. No strong active medicines of any kind should be used in this disease. The bowels should be kept regulated by the use of mild cathartics, such as rhubarb and sulphur, equal quantities combined, taken daily in a sufficient quantity to produce from two to three stools a day. Dr. Wright recommends a tea spoonful of common salt taken in water every morning for this purpose. It the disease is attended with great debility, a chalybeate pill may be taken night and morning--or take a decoction of burdock-root, sarsaparilla and wild cucumber once or twice a day, say a gill twice a day for an adult; by these means the morbid matter may be thrown off, the fluids corrected and a healthy and vigorous action imparted to the system. Wash the tumors with casteel soapsuds, and then anoint them with cedar oil, then apply the powders of ever-green plantain. When the ulcer is deep, you should use some stimulating wash after the soapsuds--such as a decoction of bayberry, wild lettuce, dewberry, brier-root, witch hazle, beach-bark or leaves, or spice-wood, after which apply the oil and powders.--The tumors should be dressed in the above manner every twelve hours. When the inflammation ceases, the use of the powders may be discontinued and healing salve applied in its stead. Before the tumor breaks, an ointment made by stewing together two-thirds fresh butter to one-third cedar oil will answer much better than cedar oil used alone. The diet and drink should be of a light and cooling nature, such as good light bread with tea, coffee or milk, soup of the flesh of young animals well prepared, with an occasional glass of wine. Moderate exercise should be regularly taken. Cold and damp should be particularly guarded against. This disease often afflicts persons for years, the ulcers extend to the bone, and a very offensive matter is discharged. For ulcers of this last kind, in addition to the above treatment, look under the following head--Ulcers--for additional remedies.

ULCERS.--(Yah-nah-wa-skur.)

By Ulcers, is commonly understood an old running sore, and it is in this sense that we here use this term. Sometimes caused by slight wounds or bruises. At other times they appear to be constitutional or a hereditary disease in connexion with a scrofulous habit. These latter are generally tedious and slow to heal.

TREATMENT.--The ulcer should first be well washed with casteel soap suds, next bathe the part in a strong decoction of beach bark or leaves, after the part is well bathed, dry it perfectly dry and anoint it with cedar oil, and apply a poultice made by thickening rye meal or wheat bran in a strong decoction of black-oak bark, the face of the poultice should be smeared with a little cream or lard to prevent it from sticking. If this treatment should not allay the fever and reduce the swelling in a few days, apply a poultice of polk-root and may-apple root, boiled to a strong decoction, thickened and applied as above; this last poultice is to draw out any offensive matter that may be lurking at the bone, and must be continued until the inflammation subsides--after the fever a bates, the black-oak poultice may be re-applied during the whole time, the wound must be regularly washed, bathed and anointed as above directed, every twelve hours. The patient should occasionally cleanse the bowels with anti-billious pills, or some other cathartic, and make a constant use of a decoction of white sarsaparilla and wild mercury to cleanse and purify the blood. The sarsparilla and wild mercury may be taken in powders or pills if prepared. The diet should be light and nourishing, every thing of a stimulating or heating nature must be avoided, particularly ardent spirits. Charcoal applied by sprinkling it on the poultice, will cleanse or purify the sores and prevent them from having a disagreeable smell. A salve made of Jamestown weed (Jimson) is very good for tedious ulcers, as is also a salve of alder-bark.

CANCER.--[Oh-tah-yeh-sku.]

The term Cancer had been applied indiscriminately to all eating, spreading ulcers; of a virulent kind. Of the cancerous ulcer, there appears to be several kinds; but the medical profession have reserved the term cancer for the most malignant and incurable kind. The appearance of the real Cancer is as follows: It commences with a small inflamed pimple or lump of a bluish color, which becomes a sore with hard rising edges of a ragged appearance. On close examination of the sore, you will discover two whitish lines, crossing from the centre to the edge of the sore. At first a burning sensation is felt in the sore, which is accompanied as the disease advances with sharp, shooting pains. After some time these pains subside and the cancer discharges a very offensive matter; this discharge gradually increases and the matter communicating to the adjoining parts, finally forms a large offensive sore or ulcer, of a most dreadful and exhausting nature, always terminating (unless cured) in a lingering, painful and horrible death.

Cancers are usually seated in some gland, but are sometimes seated in some other part. They generally make their appearance about the lips, the nose and breasts, but sometimes on other parts of the body. Those who are advanced in life, are much more subject to cancerous affections than young persons, particularly if they have scrofulous constitutions, which have descended to them from their parents.

TREATMENT.--First wash the cancer with casteel soapsuds, next bathe it well with a strong decoction of red-root, then apply a salve made as follows: Take of heart-leaf-root well pulverized, sheep suet and pine rozin, equal quantities and a smaller quantity of beeswax, stew them over a slow fire until the strength of the heart-leaf-root is extracted, then strain for use. The cancer should be washed bathed and dressed in the above manner every twelve hours; but some attention should be paid to the general health of the patient, or all the external applications may fail to effect a cure. The bowels should be kept regular by the use of the anti-billious pills, or some other cathartic. The patient should drink bitters to cleanse and purify the blood, such as sarsaparilla, wild mercury, or some similar bitters, and make a free and general use of sassafras tea. The diet must be light, such as rice, chicken, squirrel or venison, cooked in their own oil alone, and salted just enough to make them palatable; strong diet of all kinds should be avoided. This disease requires time and perseverance, but I have never known the above course fail to effect a cure, when properly attended to, says Dr. Foreman.

ANOTHER MODE OF TREATMENT.--Take the green switch of yellow-root, and the moss out of the river, burn them into ashes, then take hogs lard, or mutton tallow, and mix with the ashes and apply it in the form of a plaster to the cancer. In the first stage of this disease, narrow-doc-root bruized and steeped in vinegar, is a good application.

SORE LEGS--(Oh-nuh-sco-hah.)

Sore Legs frequently arise from neglected bruises, cuts, &c. It sometimes runs in families for several generations. When it runs in families, it is generally such families as are addicted to scrofula, scurvy and similar diseases.-- This disease bears so close a resemblance to scrofula, and the treatment for it is so near the same as the treatment for that disease and ulcers, that it would be useless to write much on the subject. But as I am personally acquainted with several persons, who have been afflicted with sore legs for a number of years, I think it probable that they will more readily find and understand the mode of treatment if laid down under its proper head.

TREATMENT.--Where sore legs are of long standing, the general health of the patient must be attended to. First, give a dose of antibillious pills to cleanse the stomach and bowels and next let the patient make a constant and free use of a decoction of sarsaparilla and wild mercury, or some other articles to cleanse and purify the blood and increase the general tone and strength of the system. Wash the leg well with casteel soap, and bathe it in a strong decoction of beach-bark or leaves, next anoint it with cedar oil, as directed for Ulcer. It should be poulticed as directed for ulcer; if the smell be very offensive sprinkle charcoal over the poultice. The leg should be dressed in the above manner every twelve hours. The patient should take moderate exercise, but spend the greater portion of his time lying, as this will give the affected part greater case. A salve of the root of swamp-doc forms an excellent application to old ulcers, and a strong decoction of the same forms a good wash for tedious ulcers. Diets must be light, such as are recommended for scrofula and ulcers. The use of ardent spirits must not be indulged in, if the patient wishes his limb restored to health, for all remedies will fail where this poison is taken even in moderate quantities.

WHITE SWELLING.--(Colah-te-coh-nu-go-gee.)

Different writers give various and even contradictory accounts of this most painful disease. They attempt entertaining the reader with accounts of several kinds of White Swelling, which are distinguished according to the seat of the disorder. All this I believe to be unnecessary in this work, as I offer but one remedy. I therefore make but two directions in White Swelling, viz: The inflammatory or first stage and the chronic or second or last stage. There is no disease to which the human family is liable, that has hitherto inflicted more severe and lasting misery, than White Swelling. It has baffled the skill of the most eminent physicians, and rendered hundreds of children of the finest constitutions and greatest activity cripples for life. Dr. Wright, a physician, who has been successful in treating this disease, speaks of it in the following words: "If the patient survives the severity of the first assault, he may for many years drag out a painful and miserable existence, his masecrated body filled with sores from the crown of the head to the sole of the feet, and his sufferings so protracted, violent and agonizing, that when he dies, as he will of a hectic fever, his friends and relations, and even parents, feel comfort in the thought that death has relieved him from his miseries, and willingly consign to the tomb the mortal remains of the unhappy victim."

Male children of the most active life and best health, from three or four to fifteen or twenty years of age, are most subject to White Swellings, but both sexes may be afflicted with it from a few months old to twenty-five years old; after which age, I have never known a case to occur. Some physicians believe that all White Swellings are caused by cold. I am of the opinion that very many causes of this disease are occasioned by cold, but I think that the number occasioned by bruises are equally great.

It generally makes its attack after being overheated by violent exercise and cooling suddenly. This disease is seated on the surface of the bone, and in the periorteum or membrane which covers the bone. Although this painful disease has baffled the skill of the most eminent physicians for centuries past, unless taken at the very commencement of the disease before it could be fully known whether it was White Swelling or not, yet a simple, easy, and certain remedy abounds in our own native forest. For the discovery of this remedy we are indebted to the Cherokee Indians. It has already relieved hundreds of this torturing and painful disease, and restored them to a state of health and activity. It has never failed in their (the Cherokees) hands to give general relief in a short time.--It has been tried by myself in cases where the regular and botanic physician had each a fair trial and had failed.--Cherokee remedies had the desired effect. And I feel no hesitation in saying from personal experience, that their mode of treatment will relieve White Swelling at any stage, if perseveringly attended to.

SYMPTOMS.--Sometimes the first symptom is a violent pain in the part affected, the pain continues for several days before the patient has signs of fever, the pain increases, in some instances it is of a whitish and in others of a reddish or flesh color--as the disease increases the patient becomes feverish with loss of appetite, great thirst, and flushed face--at other times it makes its attacks with more violence, (immediately after being over-heated, and cooling suddenly) with the appearance of inflammatory fever, which if left to itself, in a few days settles on some part of the limb; the part swells rapidly, with violent pain, and in this case it frequently has a high *red color,* although it is called *white swelling.*

TREATMENT.--First steam the affected part well with spice-wood, this should be done as follows: Boil the twigs of spice-wood to a strong decoction, and place the vessel under the afflicted part, covering the limb at the same time to prevent the steam from escaping, let it remain until it is properly steamed; next anoint it with cedar oil and bathe it in with a hot iron or shovel. If it is in the first or forming stage, after it is steamed and anointed, apply a poultice made of pole-cat or cat-paw bark, this bark is to be boiled to a strong decoction, and the decoction thickened with rye meal or wheat bran. This poultice is to scatter or drive back the disease, which it will do in a few days if matter is not already formed at the bone; where matter has formed at the bone the disease must come to a head--in this case steam and anoint it as above directed, and apply a poultice made of one-third poke-root to two-thirds buckeye-root, (the bark off the roots is the part used) they are to be boiled, thickened and applied, as directed for the cat-paw or pole cat poultice. When it is sufficiently ripe, lance it deep and continue the poke and buckeye poultice until a copious discharge is produced; if this poultice should give much pain it may be changed for one made by boiling the root of highland fern and thickening it as directed for the above poultices; but whenever the inflammation increases, and the part swells, the poke and buckeye poultice must be applied for a time. After the inflammation has subsided the cat-paw poultice may be applied. The affected part must be regularly steamed, anointed and pulticed every twelve hours. Cases of long standing will require more time to effect a cure than those of short duration: but the above treatment will cure, let the case be of as long standing as it may, if properly and perseveringly attended to.

FELON OR WHITLOW--(Oo-ne-scoh-hupee.)

Felon is an inflammation of the finger or thumb, and generally confines itself to the first joint. This disease bears so strong a likeness to white swelling that I am constrained to believe that it is one and the same disease, for Fellon like white swelling, evidently has its seat on the surface of the bone, or in the *periosteum* which covers the bone, it is attended with the most exquisite pain; this pain continues, in some instances, for several days before the color or appearance of the affected thumb or finger is materially changed--but if the disease is not checked, the affected part will put on a glossy or shiny appearance. I have known this torturing malady to prevent the sufferer from sleeping, for one, two and even three weeks in succession, during which time the part was regularly attended to in the usual manner with warm poultices, &c.

TREATMENT.--On the first appearance of the Felon, the circulation should be prevented in the affected part by means of tape or similar binding; the affected thumb or finger should be pressed gently between the thumb and fore finger, then wind the tape tightly around it commencing at the extremity and winding upwards towards the hand. This bandage should remain until a cure is affected. If the bandage should increase the pain so that it cannot be born, it should be gradually loosened until it can be borne; but as immediate case is a blessing which the great Author of our being seems to have denied in this torturing little plague, it is hoped that some patience will be exercised with regard to the bandage, it may be taken off once in twenty-four hours to examine the part, but must be replaced immediately. If the Felon is so far advanced as to have formed matter, next the bone, an incision should be made with a needle or lancet to the bone, and the bandage again applied moderately tight, and a poultice of bitter herbs applied to the incision. Some physicians recommend the insertion of vegetable caustic to the bottom of the incision. It is likely the caustic would render the cure more speedy but it would be a very severe application. Where the patient will not submit to the above treatment look under the head of white swelling for another mode. I have used the black poultice with much success in driving back risings of other descriptions, and when they were too far advanced to be driven back, it causes them to come to a head sooner and with much less pain than they would otherwise do. I have often thought that this poultice might prove beneficial in case of Felon, but have never tried it. But cedar oil is the "sovereign balm" in all diseases of the bone and the membrane which covers it.

PHTHISIC OR ASTHMA.--(Tse-nah-wah-ste-skow.)

This distressing complaint has long been numbered with those that could only be mitigated, and not cured; but the Cherokees profess to be master of this disease with all its wheezing terrors. It is a spasmodic affection of the lungs, which mostly comes on by paroxysms or fits. From infancy to old age, all sexes are subject to this complaint. Many children that have been afflicted with it from a very early age appear to outgrow it about the time they arrive at the age of *puberty.* Also at this age many persons become afflicted with it who have never before had it.--Where it is not hereditary, it leaves persons at or a little after middle age, say 40 or 50. But if it be hereditary, and does not go off at the time he or she arrives at the age of man or woman, it is apt to become more aggravated in the decline of life.

Causes which excite, or bring on a spell or fit, are often owing to the peculiar states of the atmosphere. It may be too hot or too cold, too dry or too damp, it may contain too much or too little electricity, for the nature of the disease in different persons. When the body is warm and sweating, sudden cold is sure to produce it; sudden changes from dry hot to damp cool weather almost invariably produce a paroxysm of this disease on those who are addicted to it.

SYMPTOMS.--For several hours, and in some instances days, before the fit comes on, you feel a slight difficulty of breathing, a weight and fullness in the breast and stomach, bad appetite and sometimes a great craving for food; headache, depression of spirits amounting to melancholy, restless feelings accompanied with drowsiness; the fit or paroxysm usually comes on of an evening or night, with great difficulty in breathing, attended almost to suffocation, a wheezing noise is made in breathing, attended by a hard dry cough at first, which gradually diminishes in toughness until a white, stringy, tough mucous is discharged from the throat and mouth, accompanied by a gentle moisture of the skin, and in some instances it amounts to copious sweating, severe palpitation of the heart, fever and sometimes vomiting attend it.

TREATMENT.--Take a half pound of garlic, three or four pods of red pepper and a table spoonful of common salt, pulverize and mix them well together, and take what will make two pills morning and night, and a greater quantity if the urgency of the case requires it; but this amount should be regularly taken when the patient is apparently free from the disease. Secondly, boil sour-wood bark or leaves to a strong decoction, then strain the decoction and boil it down to the consistency of molasses, then take common brown sugar and heat it in an oven over a slow fire until it melts and again becomes dry and lumpy, then add them together--proportions, four table spoonfuls of the molasses to one pound of sugar, to which add three table spoonfuls of sweet or British oil, put it again over a slow fire and mix it well together, and bottle for use. Give a tea spoonful of this syrup or mixture morning and evening. This last preparation of itself often effects an entire cure, but I prefer using them together as above directed.--They should be taken regular even when the patient appears to be entirely free from Phthisic. Lobelia is also very good for this complaint, either the green or dry herb digested in spirits long enough to extract the strength; take of this tincture just enough to produce slight nausea, say three times a day, this must be continued for a length of time after all symptoms of the disease have disappeared. Smoking the root or seeds of the Jamestown (Jimson) weed is also very good for Asthma. Persons afflicted with this disease suffers very much from extreme weakness and palpitation of the heart, particularly of a morning--to relieve this, swallow a raw egg every morning with a few swallows of fresh spring water for several mornings, say ten or twelve, then omit a few and again use the egg. The Asthmatic should rise early, take moderate exercise in the open air, but should avoid wet and damp. The diet should be light, nourishing and frequent. In cases where the difficulty of breathing is extremely great, temporary relief may be had by stewing together equal quantities of sage and honey, and letting the patient swallow it in tea spoonful doses. I believe bleeding to be very injurious in this disease, although it is recommended in the writings of several eminent physicians.

FEVERS.

GENERAL REMARKS.

Fever shows itself in so many ways and forms, that it is almost impossible to describe it correctly. To judge of its presence with any degree of correctness, we will have to pay particular attention to the following appearances and indications. The state of the pulse, the skin, the stomach and bowels, the breathing, the appetite, the color of the face, the change of feature, the tongue, eyes, &c.--There is generally soreness over the whole body, as if with fatigue after a hard day's labor, great thirst, violent pain in the head or back, or both, sometimes there is a constant desire to sleep, and sometimes great restlessness, sometimes the strength is greatly in creased by Fever.

From an early period, down to the present day, Fever and febrile diseases, have been the fruitful theme of speculation. The most distinguished medical men have differed in opinion as to the cause of Fevers. Theory has been piled upon theory, and the subject yet appears much in the dark. The opinion that most diseases originate from the stomach, appears to be supported by as sound reasoning, and good judgment, as any other theory that has yet been advanced. The first impression is made on the stomach by medicine, which acts immediately by sympathy. It is the general reservoir which receives those medical remedies by which the disease is to be subdued: The effects produced on the system by remedies taken into the stomach, show at once the great sympathy between the stomach and the whole system, and also the many evil consequences that must evidently follow a derangement of this reservoir or work-shop.

The principal secret of medicine is to discover the first cause of disease, and in the next place to apply suitable remedies in a proper way, and at a proper time. There is not so much difficulty in the science and practice of medicine as a great many persons imagine, if you will but attend to the causes of diseases, and watch the effects of the remedies. The fact is that any person possessing common sense and judgment, who will take their seat at the bed-side of the sick, ascertain how and when he was taken sick, and all the particulars as to the pains first complained of, and be minute in examining into the habits of the patient, will in nine cases out of ten succeed in relieving the patient, when mere theorists who prescribe for the names of diseases, without a due portion of sound judgment and practical knowledge will fail. The inhabitants of very few countries are entirely exempt from the attacks of Fever. I will therefore describe plainly the symptoms of such Fevers as are most common amongst us, so that with a little care and common judgment, the reader will be enabled to distinguish between Fevers and other diseases, and also to ascertain the exciting cause, which knowledge will enable him at once to strike at the root of the disease.

AGUE AND FEVER, OR INTERMTTENT FEVER.
(Oo-nah-wah-stee-skee.)

This disease generally makes its attack in the fall season of the year, and those who live on water courses, or on low marshy countries or situations are most subject to its attacks.

Agues are generally distinguished by names expressive of the periods of intermission or lapse of time between the fits.

That returning every twenty-four hours, is called by Doctors, *Quotidian*.
That which returns every forty-eight hours, or every other day is called *Tertians*.
And that returning every third day is called *Quartans*.

The above divisions are given in order to describe the disease more plainly, for the treatment is the same, differing only in the mildness or severity of the attack; if very severe the remedies employed should be active and powerful, on the contrary if the attack be mild and gentle, remedies less active and powerful will answer. Each paroxism or fit of this fever, is divided into three different stages: the cold, the hot, and the sweating stage. The cold stage commonly commences with a feeling of languor, debility or weakness, and an aversion to motion.--Frequent stretching and yawning; the feet and hands become cold, the skin looks shriveled, and a numbness or want of feeling is experienced over the whole body, and finally a chill comes on accompanied by a shivering or shaking, which lasts fifteen or twenty minutes and sometimes a much longer time. The pulse is small and frequent, and often irregular. As the chill or first stage goes off, the second or hot stage comes on, with a sense of heat over the whole body; the face becomes red, the skin dry and hot, increased thirst, pain in the back and head, throbbing in the temples, accompanied with great anxiety and restlessness; the respiration becomes fuller and freer but is still frequent; the pulse becomes more regular, hard and full; the tongue furred. If the attack be severe and the blood determines to the head, delirium often takes place for a time. In the commencement of the third and last stage, the intense heat begins to subside, moisture begins to break out on the forehead and generally extends itself over the whole body, the thirst diminishes, breathing becomes more free and full, and most of the functionaries resume their ordinary state and operation, but the patient is left in a state of extreme weakness.

TREATMENT.--First give an emetic to cleanse the stomach, see emetics in the Dispensatory, next give a purge, antibilious pills or some other cathartic. After the stomach and bowels have been well cleansed, give a sweat of seneka snake-root, black snake-root or burnt whiskey and red pepper, as either will answer. The sweat should be given about an hour and a half before the expected return of the chill. The bowels should be kept regulated by the use of cathartics, the above pill is preferable to any cathartic in my knowledge for this purpose. After the sweat has been given the patient should drink daily of the tonic bitters. See Dispensatory.

If this bitter cannot be conveniently had, the patient may take a strong decoction made of equal quantities of wild-cherry tree bark, the bark of the root of red dog-wood, and the bark of the root of the yellow or swamp poplar.--A table spoonful of this decoction should be taken regularly every hour when there is no fever, but when there is fever, it should be omitted. The Ague pill is also a valuable remedy for Ague and Fever. For the mode of preparing and administering these pills, refer to the Dispensatory.

BILIOUS FEVER--(Remittent Fever.)

Bilious Fever is Ague and Fever just described, under something of a different modification. In Ague and Fever there is an entire intermission or stoppage of the disease, whereas, in Bilious or Remittent Fever, there is nothing more than an abatement of the fever for a time. It has received the popular name of Bilious Fever because in most cases there appears to be an increased secretion of bile. Bilious attacks are most frequent in the latter part of summer and in the fall. It is most commonly met with along streams, in the vicinity of marshes and near stagnant waters. In warm climates, where great heat and heavy rains rapidly succeed each other, Remittent Fevers of a very malignant character sometimes prevail as an epidemic.

The symptoms, are a sense of heaviness and languor, pain in the head and back; in most cases the patient is taken with a chill, which is succeeded by intense heat over the whole body--the pulse is more tense and full than in Ague and Fever. If the attack of

Bilious Fever be severe, the eyes and skin often appear yellow, the tongue is covered with a brownish fur, the bowels are generally costive and the urine high colored.

TREATMENT.--In mild attacks of this disease, if taken at the commencement, it may be thrown off by cleansing the bowels with antibilious pills or some other cathartic, and drinking freely of some sweating tea. But if the attack be violent, more active means must be employed. Give an emetic of gulver and ipecac or Indian physic to cleanse the stomach and render its operation fully effective by giving warm water or chamomile tea freely; when the stomach is well cleansed give water gruel to determine it to the bowels. It the emetic is taken in the morning, the patient should take a dose of antibilious pills at night, say from eight to ten hours after the emetic. After the stomach and bowels have been cleansed in the above manner, give a sweat of Seneka or black snake-root; a tea of rag-weed will answer admirably and will produce copious sweating, in many instances, where nothing else will. If the fever should rise very high and produce delirium, put the hands in cold water, or rub the hands and wrists with cloths wet with cold water and vinegar, and apply cloths wet with cold water and vinegar to the forehead and temples, and bleed freely according to the strength and constitution of the patient. Care must be taken to continue the use of purges, until the stools assume a healthy appearance. Next take three table-spoonfuls of the powders of boneset, and pour on them a quart of boiling water; of this tea, let the patient drink a half-pint a day, unless it should produce vomiting, in which case the quantity must be diminished; this tea is intended to act mildly on the bowels, and also on the liver. It is one of the best correctors of the bile now known. After the disease is checked, if the patient suffers from debility, give tonics, such as dog-wood-bark, columbo-root, wild cherry-tree bark, &c. See tonics in the Dispensatory, and also in the Materia Medica, at which places you will find directions for preparing and administering medicines of this class.

Puking, purging and bleeding, is often followed to such an extent as to bring the patient to an alarming state of debility as soon as the fever leaves; in cases of this kind give the chalybeate pill morning, noon and night. Dose in this case is one pill about the size of a summer grape; this medicine acts more like a charm in extreme debility than any thing else.

I omitted stating in the proper place, that plasters of mustard-seeds, or poke-root poultices, or some other of the articles recommended for that purpose in Materia Medica,
should be applied to the feet to produce a revulsion from the head, where the fever is very high and the determination to the brain very great. This will greatly aid the bleeding and cold applications to the head and wrists in giving relief.

NERVOUS FEVER--(Typhus Fever.)
[Gah-lah-a-lee-oo-lee.]

This Fever as its name imports, affects the whole nervous system, and produces a tremulous motion of the body and limbs, and extreme debility, which attends it from the very beginning; the system appears to be sinking, great weariness, loss of appetite, low spirits, frequent sighing, are among the first symptoms; these are soon followed by dryness of the mouth, quick low pulse, and sometimes an unnatural perspiration or sweat breaks out on the skin for a time. The sleep is very much disturbed and unrefreshing, the countenance sinks, or seems to change from its natural expression of feature to a ghastly appearance, the tongue, teeth and gums soon become covered with a dark buff-colored scurf, the sight of food is unpleasant, and sometimes disgusting, from the extreme debility or weakness of the stomach. The difficulty of breathing becomes very considerable, sometimes the hands and feet are glowing with heat, whilst the forehead is covered with sweat; sometimes it comes very suddenly with chills and flushes, and at other times it is ten or twelve days, or even longer, before it shows symptoms of violence, making its advances so slowly and gradually as to produce no alarm. The following symptoms may be considered very dangerous: A constant inclination to throw off the cover; a changing of the voice from its usual tone; great weakness; picking at the bed-clothing; inability to retain or hold the urine; involuntary discharge from the bowels; slight aberrations of the mind, muttering as if speaking to one's self; a wild and fixed look, as if the eyes were riveted on some particular object. When these last symptoms occur, there is little to expect but that the case will terminate fatally.

TREATMENT.--Give an emetic to cleanse the stomach.--Ipecac, or Indian physic, and gulver, (see Dispensatory,) in, ten or twelve hours after the operation of the emetic, give a purge of anti-billious pills, or some other purge. See cathartics in the Dispensatory.

After the stomach and bowels have been cleansed in the above manner, give the hepatic pill night and morning; 4 for a dose. A gentle perspiration should be kept up by the use of snake-root tea. After the stomach and bowels have been cleansed, as above directed, the patient should make constant use of bitters of gulver, ipecac or Indian physic and bone-set, infused in good whiskey or wine.--Where there is trembling of the limbs and great agitation of the nerves, give nerve powders in the tea taken by the patient, freely. If the bowels incline to be costive, use injections of thin gruel made tolerably salt, to which add nerve powders freely. In the low stage of this disease, use wine freely: if the pulse is low, and the extremities cold, apply mustard seed plasters to the feet, ankles and wrists; also bathe and rub them well with whiskey and red pepper.

Diets must be light and nourishing, taken frequently and in small quantities. Slippery-elm tea or mucilage, is a valuable drink in this Fever; dried apple or peach syrrup or tea is also very good. All possible pains should be taken to keep the mind of the patient composed.

Tonics may be used freely in the advanced stages of this disease with good effects; camphor will have a good effect combined with dog-wood or wild cherry-tree bark. Wine or French brandy may be taken freely.

Bleeding in Nervous Fever is almost certain death, and should never be practiced at any stage of the disease.

YELLOW FEVER.

SYMPTOMS.--An attack of Yellow Fever is sometimes preceded by loss of appetite, disagreeable taste in the mouth, heat in the stomach, pain or giddiness in the head, costiveness, languor, debility, and dejection of spirits. At other times it attacks suddenly, with a chill, a pain in the head and eye-balls, flushing of the face, excessive thirst, and great prostration of strength; the stomach is very irritable, throwing up whatever is taken into it; the tongue is covered with a dark colored fur, the skin dry and hot, pulse small, and quick, and hard, the urine high colored and voided in small quantities; the eyes, and skin about the face, neck, and breast, becomes yellow; a dark colored matter resembling coffee grounds is at length thrown up from the stomach, called the black vomit. Sometimes diarrhæa takes place, the stools have a very offensive smell, and a black or green color; sometimes the victim of this disease sinks into a sleepy state and dies without a struggle. At other times putrid symptoms of a very violent character occur, and the patient dies in convulsions.

TREATMENT.--This Fever requires prompt and efficient treatment at the very onset. As soon as Yellow Fever is discovered, cleans the stomach, by giving an emetic; for this purpose the roasted root of prickly-sumac is probably the best article in my knowledge. Take a double handful of the roasted root, steep it in water until the strength is extracted, then give a half pint of the tea or more, and repeat in fifteen minutes if the first should not operate; give warm water freely to render the operation fully effective. After the operation of the emetic, give anti-billious pills to cleans the bowels--for dose see Dispensatory.--Then give a tea of barberry-root or the root of golden-seal, one or both; this tea is made by putting one-fourth ounce of the dried root into a quart of boiling water--of this the patient should drink a pint a day, or more if the stomach will bear it. The patient should be kept in a gentle perspiration by the use of seneca or black-snake root tea.--The bowels should be kept regulated through the whole course by the use of anti-billious pills, taken in small doses morning and night, just sufficient to produce the desired effect on the bowels, which must be judged of by he who administers.

SCARLET FEVER.

This disease often prevails as an epidemic, and is most frequent in the Fall season of the year, though it may occur at any season of the year. It is most apt to attack children and young persons, yet it sometimes attacks whole families and persons of all ages.

SYMPTOMS.--This, like other Fevers, commences with languor, lassitude, chills, heat, dry skin, nausea, and sometimes vomiting. The pulse is quick though languid, the respiration difficult and hurried, the skin is red like scarlet, and if the disease is suffered to progress, spots of a vivid red color make their appearance on the face and neck, and gradually extend over the whole body; the throat becomes sore, the voice hoarse, and the breathing very hurried and difficult; in the evening the fever is highest and the spots brightest. In the more malignant form of this disease, all the above symptoms are aggravated, inflamation and ulceration of the tonsils takes place, the redness or efflorescence spreads over the whole body, with appearance of swelling, the tongue, which in the milder form of the disease is covered with a white thick fur, is in the more malignant form covered with a black or brownish fur or incrustation. Derangement of the mind is common to each form of Scarlet Fever. When there is a tendency to putrifaction the pulse becomes small indistiact and irregular; the sores in the mouth and nose and throat become very putrid, and a briny substance or matter is often discharged from the nose, which takes the skin as it passes. This last stage of the disease is considered very dangerous, and if immediate relief is not had the system sinks into a state of collapse.

TREATMENT.--First give an emetic or puke to cleanse the stomach. A tea made of the bark of the shell-bark hickory is the best emetic with which I am acquainted; for this purpose it should be steeped in water until the strength is extracted, and the tea administered freely until it operates. When it is not convenient to give the above emetic, any other good emetic will answer--see emetics in Dispensatory. The bowels should next be cleansed by giving a dose of anti-bilious pills or some other cathartic. After the stomach and bowels have been cleansed as above directed, the patient should take the hepatic pill every morning on a fasting stomach--three for a dose for an adult; also give sweating teas freely, such as black or seneca snake-root; during the whole course must be kept regulated by the use of anti-bilions pills or some other cathartic, taken in small portions night and morning.

If the Fever should continue high and the thirst be great, the emetic should be repeated; if the throat should become very sore, treat it as directed for Malignant Sore Throat. If the patient should sink into extreme weakners or debility, give dogwood-root bark in tea or decoction freely to drink and by injections; also give the chalybeate pills night and morning. A little sulphur should be added to the diaphoretic tea which will prevent heart sickness, and aid in driving the efflorescence to the surface, which is the principal object in this disease. Flannel wet with spirits, may be kept to the neck, and the patient may inhale the steam of vinegar from the spout of a coffee-pot. The drink should be warm and the diet light.

INFLAMMATION OF THE BRAIN.

SYMPTOMS.--Violent pain in the head, the eyes are red, inflamed, and unable to bear the light; watchfulness, frightful dreams, great anxiety and indistinct recollection. The face becomes flushed, the skin dry, the bowels costive, the urine scant, and there is an extreme susceptibility of the whole nervous system. The pulse is irregular and tremulous, or strong and hard; the arteries of the temples and neck usually throb and beat violently. In the first stages of this disease the patient dislikes to talk, but as the disease advances, the eyes assume a great brightness, the patient becomes furious and talks wildly, and generally on subjects which have left deep impressions on the mind when in health. The tongue becomes dry and rough and of a black or dark yellow color.

Favorable symptoms are copious perspiration, discharge of blood from the nose, a dysentary and plentiful evacuation of urine. Unfavorable symptoms, are starting of the nerves, total deprivation of sleep, retention of urine, continual spitting and grinding the teeth, and violent fits of delirium.

Inflammation of the brain is sometimes a primary ideophatic disease, but it is often only symptomatic of some other complaint. Inflammation of the brain, may at all times be considered an extremely dangerous disease, and one which must inevitably prove fatal without the immediate use of active, prompt and efficient remedies.

Causes likely to produce this disease are, excessive use of ardent spirits, indulgence in violent fits of passion, intense study, excessive venery, violence about the head, as blows, &c. Sudden cold, long exposure to the direct rays of the hot sun, fevers, small-pox, mumps, and also by the sudden suppression of accustomed evacuations, whether natural or artificial. When the disease is long protracted, it often terminates in insanity.

TREATMENT.--First take blood from the arm by a large orifice or opening; wait a little while and again take blood, until a gentle sweat is seen, or the patient feels like fainting. Let the patient be kept in bed, with the head placed on high pillowing and the body in as upright a posture as possible, so as to lessen as far as possible the determination or flowing of the blood to the head. Give a dose of antibilious pills, and aid their operation by the frequent use of laxative injections. But while the above means are in use for the evacuation of the contents of the bowels, lose no time in shaving the hair from the head, and apply cloths wet in the coldest water and vinegar that can be had constantly over the naked head; if ice can be had, put a portion with the vinegar and it will be so much the better. If this does not cause the violence of the symptoms to abate in a short time, and the strength of the patient will bear it, give an emetic of ipecac and gulver (see Dispensatory,) or Indian physic, and draw a blister on the back part of the head and neck, also bathe the feet in warm water and apply plasters of mustard-seeds to them and the ankles in order to produce a revulsion of the blood from the head. I neglected to state in the proper place, that after the bleeding and evacuation of the contents of the bowels, every means should be used to produce sweating, such as the free use of diaphoretic teas or powders; for this purpose a tea of seneca snake-root and black snake-root is preferable, but where neither of these can be had, other diaphoretics will answer. After the violent symptoms abate, still keep the bowels open for several days with cream of tartar, salts, senna, or something that causes copious watery discharges from the bowels. During the whole time the room of the patient should be kept perfectly cool and as dark as convenient, nor ought the least noise to be permitted to disturb the quiet of the patient.--When the fever begins to subside and the reason to return, these instructions should be particularly attended to--because the slightest cause will, in many instances, bring on a return of the disease and with redoubled violence, which will in all probability prove fatal in a short time. Diets for several days should be of the lightest kind.

INFLAMMATION OF THE STOMACH.

Inflammation of the Stomach may proceed from obstructed perspiration, from receiving blows or wounds in the region of the stomach, from severe and frequent emetics, drinking extremely cold water when the body is over heated from corrosive poisons taken into the stomach; also by the transition of the gout, or acute rheumatism to the stomach.

SYMPTOMS.--Inflammation of the Stomach can easily be distinguished from other diseases by its peculiar symptoms, it is therefore impossible to mistake it for any other disease if proper attention be paid to those symptoms. It is always attended with a violent pain in the stomach, which is greatly increased by pressure over the stomach; there is also a burning heat at the pit of the stomach, frequent retching as if to vomit; when anything is taken into the stomach it is immediately vomited up; there is great loss of strength, excessive thirst and uneasiness, continued moving of the body from side to side. If the disease be not checked, it rapidly advances, the hands and feet become cold, the bowels costive, the countenance haggard and wears an indiscribably anxious aspect, hickups ensue and the patient soon dies.

TREATMENT.--This is a very dangerous complaint, and may terminate fatally in twenty-four hours, if not arrested in its progress. First bleed freely from the arm--the pulse is frequently low, and small, but this should not deter you from bleeding, the pulse will frequently rise on bleeding several times. Employ some anti-emetic, to check the vometing, the contents of the bowels should be evacuated as speedily as possible, by the use of laxative injections, such as gulver syrup in a strong tea of catnip. As soon as the inflammatory symptoms have been subdued by frequent bleeding, and clystering, the patient should be put into the warm bath and remain there as long as possible; as soon as he is taken from the bath apply a blister over the region of the stomach, or a large plaster of ground mustard-seeds wet with strong vinegar will answer. By turning to Dispensatory you will find directions for preparing several plasters, which will draw blisters. Small quantities of sweet oil, given occasionally will aid in allaying the inflammation. The bowels must be kept open by injections made of flaxseed tea, chicken water, slippery-elm tea or thin gruel. These injections will also aid in nourishing the patient, as it will be improper to take nourishment into the stomach. The patients drink should consist of slippery-elm or flaxseed tea taken a little below blood heat. When the inflammatory symptoms have subsided and the stomach will bear it, nourishment may be taken, but it must be done with great caution, and in very small quantities; it should consist of slippery-elm tea, flax-seed tea, new milk boiled and thickened a little, rice, light soups, or thin gruel with a little new milk in it; anything taken into the stomach should neither be too, warm or too cool, a little below blood heat is probably the safest temperature. When this disease is produced by poisons taken into the stomach, the poison must be removed by an emetic, as directed under that head, and the disease then treated as above directed. When inflammation of the stomach terminates fatally, it always ends in mortification. A sudden change from severe misery to perfect case, is conclusive evidence that mortification has taken place. Inflammation of the stomach, sometimes occurs in putrid diseases. It is discovered by inflammatory appearances on the inside of the mouth. When the face and mouth is so affected in Fevers, it is to be feared that it reaches the stomach, especially if the stomach shows unusual susceptability, accompanied with frequent vomiting--in this the progress of the disease may be arrested by giving a tea-spoonful of the spirits of turpentine in a little water. This medicine is well calculated to prevent gangrene and mortification, and must be given more or less frequently, according to the urgency of the symptoms.

INFLAMMATION OF THE INTESTINES.

In this disease the danger of mortification is great and relief must be given soon or every effort will be vain.--Symptoms are, severe griping or pain in the bowels, especially near the naval. It may easily be distinguished from inflammation of the stomach, by the pain being lower down, and from cholic, by the smallness and quickness of the pulse, and by the extreme tenderness of the belly, the pain being increased by pressure upon it; whilst in cholic it affords relief, it is attended with sickness at the stomach and vomiting, obstinate costiveness and considerable fever. Great weakness attends this disease, the urine is high colored and voided with difficulty.

Inflammation of the Intestines is produced by nearly the same causes as those which produce inflammation of the stomach. It also arises in some instances from hard indigestible food remaining in the bowels; from severe cholic, blows and wounds about the region of the bowels--by long and severe dysentery, worms, and in some instances from hernia or rupture.

TREATMENT.--In the first, stage of this very dangerous disease, it will be necessary to bleed and repeat as often as the urgency of the case requires it. A free and thorough discharge should be produced from the bowels by means of laxative injections repeated at short intervals until the desired object is obtained. A part of each injection should be composed of slippery-elm or flaxseed tea and a strong tea of cat-nip. The whole abdomen should be kept well bathed with a strong tea of catnip, red pepper and vinegar; the warm bath will be of service, but the temperature should be very moderate. It the disease should prove obstinate or unyielding, apply a blister over the belly. Sweet oil may be given in table-spoonful doses and repeated frequently; after the bowels are thoroughly cleansed, injections of catnip and slippery-elm tea must be given frequently for several days, to which may be added laudanum, about forty or fifty drops for a grown person and less for children. Purgatives in this complaint only tend to aggravate the symptoms. Your principal reliance is therefore to be placed on injections and copious bleeding, especially if the patient is of full habit. Charcoal taken by the mouth and administered by injection, seldom fails to produce good effects. The slightest causes are exceedingly apt to produce a relapse of this disease, for this reason, exposure to cold should be carefully avoided, and indigestible food should not be used; the bowels should also be kept regulated by the use of mild and cooling injections. Diet should be very light and taken in very small quantities, and nothing better than slippery-elm tea could be recommended for the patient's constant drink.

INFLAMMATION OF THE KIDNEYS.

SYMPTOMS.--A severe pain about the small of the back, some degree of fever, the pain frequently darting down the thigh or thighs, as the case may be. The urine voided in small quantities and with difficulty, of a pale or reddish color. The pain in this disease is seated nearer the back bone and loins than in cholic. The bowels are costive, the skin is dry and hot, there is nausea and sometimes vomiting. The slightest motion or jolting gives great pain, and even sitting upright in bed produces restlessness, the patient always experiencing the greatest ease, when lying on the affected part. Sometimes one and sometimes both of the testicles are drawn up to the belly, so that you can scarcely feel them.

The causes most likely to produce this complaint, are wounds or bruises of the kidneys, calculous concretions lodged in them, the too free use of active diuretic medicines, great exertions in lifting, violent and sudden strains, exposure to cold when overheated, and lying on the damp ground.

TREATMENT.--If the patient be stout and full of blood, the lancet should be used; the bowels must be opened by mild clysters and oil, cream of tartar or some gentle purge. A mixture of sulphur and cream of tartar is an excellent preparation for keeping the bowels open in this complaint. Flannel cloths wrung out of hot catnip tea, or a decoction of red pepper and vinegar should be applied over the small of the back. After the bowels have been cleansed as above directed, give the diuretic powders morning and night a tea spoonful for a dose, at the same time let the patient drink a tea made of the piny weed root, at least a half pint a day; if the piny powders cannot be had, a tea of the common rush will answer, but it is not so good as the piny.

All the drinks should be made warm, and a portion of peach-tree gum dissolved in them. Slippery-elm or flaxseed tea will answer a good purpose. A strong decoction of peach-tree leaves, either green or dried, is a most excellent remedy in this complaint, and has of itself affected many wonderful cures.

Diets should be light; onions, although not a light diet, will answer well in this disease, where the stomach will bear it. The patient should be kept quiet and easy, and free from cold while any appearance of inflammation continue. When the patient begins to recover, moderate exercise in the open air will be proper and advantageous.

INFLAMMATION OF THE BLADDER.

SYMPTOMS.--This disease is known by a sharp pain at the bottom of the belly, immediately above the privates, the pain is much increased by pressing on the part with the fingers; there is sometimes considerable uneasiness in the lower part of the back. There is a constant desire to make water, which is passed with much difficulty and in very small quantities, it is high colored, and not unfrequently tinged with blood. It is also attended with sickness and vomiting and a constant desire to go to stool; the bowels are bound, the pulse irregular, and always some fever. Inflammation of the bladder is produced by obstructions in the urethra, by suppression of urine, by the formation of calculous concretions, and by costiveness. It may also be produced by injuries, such as severe blows, falls, kicks, &c., by holding the urine a great length of time, and by taking the tincture of cantharides or Spanish flies.

TREATMENT.--In this disease, when the patient is of full habit and there is much fever, bleed. The bowels must be opened by cooling purges and injections, after which, the patient should take of the diuretic powders, as directed for inflammation of the kidneys, and also drink of the tea of piny powders or rush. The region over the bladder should be frequently bathed with a strong decoction of red pepper and vinegar, or a poultice of catnip applied over the part. When this complaint is caused by the lodgment of a stone in the neck of the bladder, the cause should be removed as directed under the head of Gravel, at the same time using the above means to allay the inflammation. A tea of slippery-elm or flax-seed injected into the bladder, will be found very good. Every time the patient has to make water he should sit over the steam of pine tops, cedar tops, or bitter herbs; this will greatly assist the passing off of the urine, and also in relieving the pain occasioned by voiding it. Diets and drinks of a heating nature should, by all means, be avoided.

INFLAMMATION OF THE SPLEEN.

SYMPTOMS.--In this disease there is considerable pain in the left side, just under the ends of the ribs, and round to the back-bone. In severe cases the pain reaches up to the left arm-pit and into the shoulder; the skin and eyes are yellow. The pain which extends up the side, may easily be distinguished from the plurisy, by numbness and deadness about the shoulder joint, and also by the seat of the pain being below the ends of the ribs; the symptoms most to be relied on are, puking of blood, watchfulness, great weakness, and very frequently the mind is much confused: there is also considerable fever.

TREATMENT.--Purge well with antibilious pills. (See Dispensatory for dose.) After which, they should be taken in doses sufficiently large, night and morning, to keep the bowels gently open. These pills are peculiarly adapted to this disease. The side should be bathed frequently with a strong decoction of pepper and vinegar or essence of pepper. After the inflammatory symptoms have in some degree subsided, the patient should drink bitters, composed of equal quantities of gulver-root and bone-set leaves, and a much smaller quantity of Indian physic infused in spirits. If the spirits should disagree with the patient, the hepatic pill or anti-dispetic syrup, taken night and morning will answer. In chronic cases, after bathing the side as above directed, for a few days apply a strengthening plaster. Diets must be light and nourishing, and the exercise moderate.

This complaint is brought on by long continued fevers, or by long continued fever and ague, and by affections of the liver. What are commonly termed ague cakes, are diseases of the spleen, and sometimes terminates in *Inflammation of the Spleen*.

INFLAMMATION OF THE LIVER.

There are two species or forms of this disease, distinguished as acute and chronic inflammation of the liver.

Symptoms of acute Inflammation of the Liver.--In this form of this disease, there is a severe pain in the right side, from the ribs to the hip, accompanied with fever and slight chills; the pain often rises to the point of the shoulder, and extends to the collar-bone; there is hard breathing, dry cough, a tightness across the breast, an inclination to lie on the right side and yet hard pressure on the right side increases the pain. The bowels are costive, the urine high colored, the pulse frequent and hard, and the tongue covered with a whitish fur. There is often sickness and vomiting of a bilious matter. The skin is dry and hot, and if the disease is still permitted to advance, the skin and whites of the eyes assume a yellow color.

TREATMENT.--If the inflammation is considerable, and the pain severe, bleed, and follow the bleeding by cathartics. The anti-billous pill is probably the most suitable preparation for this purpose. After the inflammatory symptoms have been subsided by bleeding and purging, give the hepatic pill night and morning--three for a dose if the stomach will bear them, if it will not, give them in such doses as the stomach will bear. The bowels must be kept open by the daily use of the anti-billous pills in small doses, the quantity being best judged of by the patient. The patient should also drink freely of a strong tea of liverwort, and use bitters composed of one-third silkweed root and two-thirds butterfly root. A tea of spicewood forms a valuable drink in this disease, as it promotes a gentle perspiration. Blistering the side often gives great relief where the pain is severe.

Chronic is a term applied to diseases which are of long continuance, and are generally attended with but little fever. Chronic affections of the liver, is commonly best known by the name of *"Fever Complaint"* It may either be a consequence of the above, or it may come on gradually, without acute inflammation. The chronic form of this complaint is generally produced by exposure to sudden viscissitudes of heat and cold, by the intemperate use of spirituous liquors, by long continued attacks of intermittent and remittent fevers, and by the improper treatment of measles and other diseases.

SYMPTOMS.--Chronic inflammation of the liver, is frequently so mild at its commencement, and so very obscure in its attack, as to produce but little pain, and excite but little uneasiness, until the disease is firmly seated, at which stage it is tedious to cure, and if cured, requires a persevering use of the remedies, with the most scrupulous attention to regimen and diet. It is attended with general weakness and dislike to motion, indigestion, flatulency or frequent belching of wind from the stomach, a short dry cough, and occasionally, difficulty in breathing. The bottoms of the hands and feet are generally dry and hot, tho' sometimes moist and cold. A dull pain or misery is felt between the ribs and right hip, extending at times to the right shoulder. The bowels are mostly bound, but sometimes become very laxative for a few days--the stools are generally of a clay color, and occasionally particles of blood are seen among them. Whatever is taken into the stomach as food, frequently sours, and produces pains in the stomach, and an acid taste in the mouth and throat.-- There is often a burning at the stomach, somewhat different from heart-burn, and an unpleasant headache, with frequent giddiness or swimming. The urine is high colored and usually scant, the complexion and countenance assumes a sallow or diseased appearance, and the whole system is oppressed with an unusual sense of fullness.

TREATMENT.--First purge with anti-billous pills, butternut syrrup, or black-root pills or syrrup, then give the hepatic pill, three every morning, if the stomach will bear them, and two, if three cannot be retained without producing great nausea. A chalybeate pill should be taken every night about the size of a pea or summer grape.--The patient should also drink freely of the tea of liverwort, and bitters, as directed for *Acute Inflammation of the Liver*. During the whole course, the bowels must be kept open by the use of cathartics. The Dispensatory shows several valuable preparations for this purpose. Bathing the feet in warm water frequently, will be found of service, or if convenient, the warm bath is much better.

The diets must be light and taken in small quantities, moderate exercise on horseback or otherwise, will be necessary. But all cold, immoderate exercise and exposure of every kind must be avoided, if a cure is desired.

CONSUMPTION.
(Oo-coh-yoh-ter-tsu-ne-si-wah-skan.)

This complaint is marked by a general wasting of the body; great weakness is felt on the slightest bodily exertion; the pulse is quicker than natural, small and irregular; a short dry cough which becomes more troublesome at night; a white frothy mucous is spit up. As the disease advances, a pain, and sensation of heat and oppression is felt through the breast, extending up to the points of the shoulders, the spitting becomes more copious and frequent, and is sometimes streaked with blood--sometimes it is dark, and at other times it is of a yellow or green color, having a remarkably unpleasant smell; when put into pure water it sinks to the bottom, while common mucous floats on the surface of the water; the urine is high colored, and deposits a muddy sediment, the cheek or cheeks frequently flush with hectic fever, which lasts one or two hours, and then gradually goes off; the palms of the hands and the soles of the feet are mostly hot; the pulse gradually grows quick and hard--these symptoms are soon followed by profuse night sweats. In the last stage of Consumption the countenance assumes a ghastly, unnatural appearance; the voice becomes hoarse, hollow and unnatural; the white part of the eyes have a shiny or pearly appearance, while the eye itself beams with uncommon lustre; the nails are of a purple color; there is frequent purging, and great difficulty in breathing, amounting at times, almost to suffocation. When these last symptoms occur, the case may be considered desperate.

Obstructions from cold in some way or other, is the common cause of Consumptions. It is most apt to attack persons between the age of twelve and thirty; but it sometimes attacks persons at the age of fifty. In youth when a change of voice takes place and the lad enters the incipient stage of manhood, there is considerable debility experienced, and not unfrequently accompanied with a short dry cough. This is a critical period, and a little carelessness or neglect may end in an incurable attack of Consumption. Such persons as have been raised tenderly, without due exercise or fresh air, will be much more liable to an attack of this ever-to-be-dreaded disease, than those who have been accustomed to daily labor or exercise. Damp air, damp beds, damp clothes, is often the cause of Consumption; it is also caused by inflammation of the lungs, suppression of the menses in females; tight lacing; diseases at the liver and stomach. It is hereditary and often takes whole families as fast as they approach man or womanhood.

TREATMENT.--The patient should commence by taking a tea spoonful of the mixture or syrup for Consumption night and morning; it is made as follows:

Take a table spoonful of tar, the same quantity of honey and the yolk of an egg, mix these articles well together; the tar should be a little warm, that they may the more easily mix. A large quantity may be prepared observing the same proportions. If the patient is very weak and the above dose operates too severe, give a smaller portion, he who administers will be best able to determine as to this. It will be three or four days before this medicine gets into full operation, and when it is taking hold of the disease to advantage, it causes the patient to expectorate or spit up mucous from the lungs with great rapidity; while taking the above, the patient should also take the inner bark of the yellow pine, and spikenard root, of each one pound, keep it constantly by the fire that it be warm, and use it as a constant drink. I have used the chalybeate pill night and morning, in connexion with the above remedies with the happiest effects. Where the patient is laboring under great debility it is without doubt one of the best stimulating and tonic medicines in the world.

Diets should be light; all kinds of rich and oily food should be avoided; buttermilk and corn or rye mush is very good, as is also rice, half cooked eggs, and milk drank warm from the cow morning and evening. Squirrel or chicken may be eaten by some, but others cannot use either without injury; the patient or administering physician will have to exercise some judgment on this subject. Wet and damp of every kind must be avoided, regular but moderate exercise taken, the mind should be kept cheerful; long journeys are spoken of by some physicians as being advantageous, but I cannot agree with them in this particular. They unavoidably produce irregular habits in eating, drinking, sleeping, using medicine and often keep the mind in a high state of anxiety about things left behind, &c., all of which produce injury rather than benefit. Regularity of habits is indispensably necessary in the cure of Consumption, I would therefore advise persons afflicted with this disease to remain at home with their friends, for I can assert that the kindness of friends in connexion with neighborhood exercise and amusements, will greatly aid in restoring health, where long journeys would only fatigue the patient and aid the disease in wearing out and extinguishing the little remaining spark of animal life.

Bleeding in Consumption is a most pernicious practice and the sooner it is abandoned the better it will be for those who are afflicted with this alarming and too often fatal disease. Yet it is recommended by a majority of the most eminent physicians of our age. Dr. Wright, however, is an exception, he disapproves bleeding in this disease in the strongest terms. He says, "The disease itself proceeds from debility, which produces obstructed perspiration, and nature not being able to relieve the lungs from the matter thrown upon them, acts as an irritant and occasions coughing and diarrhæ, and in proportion to the vital fluid you abstract, you impair the strength of the patient, and open a road for the incursions of the enemy." Dr. Wright has been more successful in the treatment of Consumption, than any physician among the whites, with whom I have ever had an acquaintance, he in many instances restored persons to health after they had tried such remedies as are usually prescribed for Consumption, and had been given over as incurable, his treatment is simple, and consists of remedies that may be procured by any person. I give it in full that those who wish to try it may have it in their power to do so.

DR. WRIGHT'S TREATMENT FOR CONSUMPTION.

Take his chalybeate pill night and morning, and through the day drink from a pint to a pint and a half of Dr. Wright's beer for Consumption. This course should be pursued with regularity.

Diet should be light and nourishing, such as butter-milk and rye mush, half done eggs and the like, he also recommends new milk of a morning.

The success of Dr. Wright in the treatment of this disease, is acknowledged by all those who were acquainted with his practice.

RUPTURE OR HERNIA.

Rupture or Hernia is an unnatural protrusion of a portion of the bowels or intestines, through the lacerated fibers or muscles of the part, where the swelling occurs.

It may be produced in children by excessive crying, coughing, vomiting, and it is frequently produced by gravel. In people who have passed the prime of life and in those who are in the full vigor of maturity, it commonly originates from extraordinary exertion, such as jumping, fighting, wrestling &c., or by violent blows or injuries about the abdomen, that lacerates the muscles without lacerating the skin.

TREATMENT.--On the first appearance of an injury of this kind the protruding portion of the intestines should be replaced, as there is great danger of the parts becoming inflamed and so enlarged that it cannot be returned, in which case there is danger of mortification. For this purpose, place the patient on his back, raising his hips higher than his head by means of pillows or bed clothes folded up; by placing the patient in this position, the protruded
part may be replaced by a gentle pressure with the fingers, if the parts be not swelled, the operation may be rendered more safe and easy by applying cloths rung out of hot water to the affected part as warm as can be borne, or a decoction of catnip applied in the same manner will answer admirably. If the parts have become inflamed and swelled, you should not attempt replacing the protruding portion, until means have been employed to reduce the inflammation, and swelling: to effect this thicken a strong decoction of rattle-root with cornmeal or flour, and apply over the affected part, this will give ease, take out the inflammation and produce relaxation, so that the protruded portion of the intestines may be returned by the hand, by placing the patient on his back as above directed; when this has been accomplished, apply over the place a plaster of red oak, which is made by boiling the bark until a strong decoction is obtained, then strain and continue boiling it until it is reduced to the consistence of thick molasses. A truss must be worn to keep the parts to their proper place, the truss should be confined by means of a broad bandage, which should extend around the patient and be kept moderately tight, it should be worn a sufficient length of time for the parts to regain their strength.

RHEUMATISM--(Tsi-tah-nah-ler-la-skah.)

This very painful disease, in which the poor sufferer drags out a miserable and wretched existence, is quite frequent in the western country. It is brought on by exposure to cold and wet, by remaining too long on the damp grounds, by sleeping in damp places or by sleeping in a free current of air at night, by exposure to dews, by changing a warm dress for a thin one, by being greatly heated and becoming suddenly cool. This complaint may occur at any season of the year when there are sudden changes from heat to cold or from wet to dry. Persons of all ages are liable to its attacks, but adults and those advanced in life, and those whose employments subject them to sudden transitions from heat to cold are most liable to its attacks. This disease is distinguished into two kinds, as acute, or inflammatory, and chronic; when both fever and inflammation accompany the pain, it is called acute or inflammatory Rheumatism, and when little or no fever and inflammation attend the pain, it is called chronic Rheumatism. There is also a disease called by physicians Rheumatic Mercuriatis, which means Rheumatism produced by the improper use of mercury, that is, by permitting the mercury to remain in the system without giving the proper remedies to carry it off, which is flour of sulpher and a free use of diaphoretic teas. Flour of sulpher is nothing more than brimstone purified and pounded to a fine flour, it is a true and certain antidote against mercury.

SYMPTOMS.--An attack of acute or inflammatory Rheumatism usually commences with chills, succeeded by heat, thirst, restlessness, anxiety, a hard, full, quick pulse, and other symptoms of inflammatory fever. Next an acute pain is felt by the patient in one or more of the large joints, followed by a tension and swelling of the affected parts. The pain often shifts from one joint to another, leaving the part previously occupied red, swollen and very tender; the tongue in most instances white; the bowels costive, and the urine high colored.

Chronic Rheumatism may either be a consequence of the termination of the inflammatory, or it may arise independent of it. When inflammatory Rheumatism terminates in the chronic, the parts, which were affected with inflammation, are left rigid, weak, and in some instances, puffed, and the pain being no longer moveable, is confined to the same parts; some instances, however, occur in which it shifts from one joint to another, but it is unaccompanied with inflammation or fever.

INFLAMMATORY RHEUMATISM.

TREATMENT.--Give bitters composed of a half pound of prickly-ash bark of the root, one-fourth pound of rattle-root, and two ounces of blue-root, digest the whole in one gallon of whiskey--of this the patient should drink freely three times a day, or what the stomach will bear. The affected joints must be regularly anointed with the rheumatic ointment--see Dispensatory.

The bowels should be cleansed and regulated by the use of anti-bilious pills.

CHRONIC RHEUMATISM.

TREATMENT.--In this disease give bitters of rattle-weed root, prickly-ash bark of the root, and prickly-sumack bark of the root an equal quantity of each, digest them in whiskey, and take what the stomach will bear three times a day; anoint the affected part with the rheumatic ointment--see Dispensatory.

The bowels should be cleansed and regulated by the use of antibilious pills or some other cathartic. In all cases of Rheumatism the patient should carefully avoid sudden changes from heat to cold from dry to wet or damp, night air, violent exercise, sudden check of perspiration and exposure of any kind. Regular but moderate exercise should be taken, and the above treatment strictly attended to.

Diets should be light and nourishing, and such as best agree with the patient.

I have witnessed with surprise and pleasure the result of Turk's mode of treatment for Rheumatism, and as it is a remedy which is in the reach of every farmer, or inhabitant of the western as well as other States, and that can be prepared and used with safety by any person who possesses five grains of common sense, I think it probable that by giving it a place in this work it may enable some fellow being to relieve him or herself of this most painful disease, who would otherwise drag out their lives in misery and wretchedness. I have never applied his remedy myself, but know the ingredients to be excellent in this complaint.

TURK'S TREATMENT FOR RHEUMATISM.

"Take one half bushel of well washed Polk-root--this root is best when dug in the winter and in the dark of the moon--extract the strength by boiling it in clear water, when you put the polk root in to boil, put in five or six pods of red pepper, when the strength is extracted, strain the decoction, and continue boiling it until it becomes very strong, then add a quart of oil, and continue boiling or simmering until the water is entirely gone. The strength of the pepper and polk root will remain in the oil after the water is extracted. When this oil (becomes cool, but while it will still run, it should be put in stone vessels, as it will eat tin up,) it should be kept closely covered, or it will mould. I have used several kinds of oil; the fish, eel, cat fish, beef leg oil, the oil of the Guinea pig, and the oil of the fat cut dogs, and have been successful with all, but the oil of the fish, Guinea pig and dog, I prefer. It will cure when fresh but I think it gets better as it gets older.-- The oil of which it is made must be pure and in no case rancid.

Mode of application.--As this disease attacks the joints only, this medicine should be rubbed all around the affeced joints and well bathed in by warming the hand frequently and rubbing the joint: a warm fire shovel will answer. If the disease is in the hip joint, apply the ointment to the whole back bone and around the hips. This ointment should be applied in the above manner, twice a day for five days, and then once a day for five days more, and if the cure is performed quit, but if the pain should return in the slightest degree, again apply the ointment. Where there is much swelling I use the bandage after putting on the ointment, if the swelling is in a place that can be bandaged.

The system must be put in good order, and kept so, that the medicine may have a fair chance, as it has a powerful enemy to contend with. I have met with cases in which I had to prepare the system for a few days before I used the ointment. In the inflammatory Rheumatism, I bleed freely, and in the chronic kind I bathe in warm water, and before putting on the ointment wash off well and let the parts dry well, as oil and water do not go well together. The quantity of oil will be best known by its going in. If several joints should be affected, put the oil on one in part, after the other, then repeat; a soreness will take place, this is caused by the rubbing, and no danger in it. I have seen some cases that seemed to get well in a few days, and then get worse; this is the time that nature is, with the aid of the medicine, fighting for power, and the result always turns out a cure. Do not stop using the medicine at this time, regardless of the number of days. The bark of burdock root, and sarsaparilla, an equal quantity of each filled in a bottle, and good spirits poured over it, must be drank freely three times a day, perhaps it might answer to put the roots into water, but I have never cured a case without the spirits."

JAUNDICE.--(Foh-lo-ne-ga-tse-nah-noh-stee.)

SYMPTOMS.--This disease is characterized by yellowness of the skin, and whites of the eyes; the urine is high colored and leaves a yellow sediment in the vessel after standing a while; the stools are clay-colored; a dull, heavy, languid feeling prevails, attended with costiveness; the pulse is sometimes strong and full, at other times weak and feeble; chillness for a time, succeeded by flushes of heat; a bitter taste in the mouth, nausea and sometimes vomiting; a restless, uneasy sensation is experienced throughout the system.

CAUSES.--This disease is always caused by some derangement of the Liver and the parts connected with it. It may be caused by anything that obstructs the passage of the bile through its natural channel; a sudden stoppage of the menses in females, or the discharge in clap. Indulgence of anxious thoughts, or of any depressing passions, the excessive use of ardent spirits, a sedentary life, &c.--When Jaundice is produced by biliary obstructions, caused by gall-stones lodged in the biliary ducts, acute pains will be felt in that region, which will be increased by eating. The pain produced by the passage of a stone along the biliary ducts, may be distinguished from the pain produced by inflammation of the liver, by the acuteness of the former. This complaint sometimes originates from inflammation or scirrosity of the liver or spleen. When it originates from the last named causes, and is suffered to run on for a length of time without medical aid, it is seldom cured. In the last stages of this complaint the skin is often marked with black spots or streaks. In some instances from fifty to a hundred and even more of these gall-stones have been taken from a dead subject on dissection, and the gall-bladder found greatly distended.

TREATMENT.--The first object should be to cleanse the stomach and bowels. For this purpose give an emetic of American ipecacuanha, or Indian physic, to either of which may be added gulver-root if prepared--from one to two tea-spoonfuls of the powdered root is a dose--if the gulver be added, give about that quantity when combined.--Give injections of gruel with a little table-salt, and hogs lard in it if the bowels be hard to move. After the emetic has operated by gentle vomiting, give gruel, which will determine it to the bowels. After the alimentary canal has been cleansed in the above manner, a strong decoction of wild cherry-tree bark should be drank freely. To relieve the pain in the side which usually attends this disease, rub the side with the essence of red pepper; also give the hepatic pill night and morning, two for a dose.

Diets should be light and nourishing, a raw egg should be taken every morning on a fasting stomach. Some persons when directed to take a raw egg, will beat it well in a saucer or other vessel, and mix it with sugar or spirits or both, this in some degree cooks the egg and destroys its medical virtues--it should be taken from the shell and swallowed in its natural state. Fruits, light bread, sour milk and mush, whey, &c., will be suitable diets in this disease. Gentle but free exercise should be taken, this will have a tendency to open the pores and restore health, and as Jaundice produces great depression of spirits, every possible means should be used in exercise, amusements, company, &c., to secure tranquility and cheerfulness of mind.

FALLING OF THE PALATE.

The elongation or falling of the palate, is attended with a tickling in the fauces and soreness at the root of the tongue. It generally proceeds from a foul stomach.

TREATMENT.--Gargle the throat frequently with some astringent tonic article, such as a strong oose of oak, persimon, blackberry brier root, &c., the gargle should be sweetened with honey; avoid speaking as much as possible. If this should not give relief, give an emetic. Alum water forms an excellent gargle. Dr. Ewel and Dr. Wright recommend the application of pepper and salt to the elongated Palate by means of a spoon-handle. I have seen the Palate restored to its proper place by tying a lock of hair on the top of the head so as to draw the skin tight.

POISONS.--(Oo-skoh-sog-tee.)

Poisons are of three kinds, as animal, mineral, and vegetable.
Animal poisons are such as are communicated by the bites of poisonous reptiles or the stings of poisonous insects.
The principal mineral poisons are the different preparations of arsenic, murcury, copper, zinc, antimony, lead, tin, &c.
The chief vegetable poisons are, henbane, night-shade, sicular or hemloc, fox-glove, wolf's bane, laurel, opium, Jamestown weed, mush-rooms, and black sarsaparilla.

ANIMAL POISONS OR SNAKE BITE.
(E-nah-tuh-oo-ne-skah-low-tsuh.)

When an individual is bitten by a poisonous serpent, if it be on any of the extremities, immediately tie a bandage or ligature around the limb, between the wound and the body, this will greatly retard the passage into the system. Give the patient a large dose of the saturated tincture of lobelia, if it can be had, if it cannot be immediately procured, bruize the lobelia herb, put it in whiskey and administer it freely, until copious vomiting is produced. If neither of the above articles can be had give some other emetic, but lobelia is far preferable to any other in my knowledge. When the emetic is done operating, give an infusion of the root of rattle-snake's master. This infusion or tea should be drank freely, as it is entirely harmless in its operations on the system. For an external application to the wound, use the bruized root of the rattle-snake's master. This treatment will cure the bite of the copperhead or rattle-snake, or any other poisonous reptile.

ANOTHER MODE OF TREATMENT.--Apply the ligature or bandage and administer the emetic as above directed, and after the operation of the emetic, give a tea of piny-weed root freely. For an external application to the wound, make a plaster to the wound of equal quantities of salt, tobacco, indigo and hogs-lard; pulverize the tobacco, indigo and salt, then mix all the articles together and apply it in form of poultice. A free use of spirits, such as whiskey brandy, &c., will be found of great benefit in all cases of bites or stings. I have ascertained from personal observation, that a person when intoxicated, cannot be poisoned by the bite of a snake. Many lives have been saved among the Indians, by the free use of whiskey and red-pepper; indeed, I believe, that whiskey alone will save life in many instances, when the bite would prove fatal if an active remedy was not resorted to immediately. The quantity of spirits taken need not give the least alarm, for I believe it to be impossible to give enough to do an injury. The pulse should be frequently examined, and whenever it begins to sink or grow feeble and fluttering the spirits should be immediately resorted to until the pulse is raised.

There are many herbs which may be used to advantage in snake bite, such as striped blood-wort; when this herb

is used, apply the bruised leaves to the wound, at the same time taking the expressed juice internally in table-spoonful doses repeated every few minutes.

INDIAN SANIDE.--When this is to be used, make a decoction of the root and give it in doses of half gill every half hour, and at the same time apply the bruised leaves to the wound.

MOUNTAIN DITANY.--Of this apply the bruised herb to the wound and drink freely of the tea.

COMMON GREEN PLANTAIN.--Bruise the herb and root and apply it to the wound, and at the same time take the expressed juice or tea freely. There are many other herbs that are good for snake bite, as may be seen under their different heads. I have known the bite of the copper-head cured in the following manner: Immediately apply to the wound, tobacco, which has been perfectly wet in vinegar, and as soon as it can be prepared give a strong decoction of the yellow-poplar root bark, and bathe the wound frequently with the same. When the bite or wound enters a large vein, the only chance to save life is to keep the stomach in motion by the use of emetics, and the pulse from sinking, by the use of whiskey or spirits of some kind, at the same time using external applications to the wound to kill or extract the poison.

STING OF INSECTS.

The sting of insects will seldom need any thing more than to wash the wound with the tincture of lobelia, or to apply the bruised leaves to the wound. Tobacco wet with vinegar, is an excellent application, or any of the articles recommended for snake-bite will answer. By applying to some of those simple means, persons may relieve themselves of severe pain, and sometimes sickness. The sting of many insects that are not dangerous, often produce great pain and disagreeable swellings. Spider bites of poisonous appearance, should be treated as snake bites.

VEGETABLE POISONS.

The symptoms which follow eating or swallowing vegetable poisons, are loss of memory, confusion, vertigo, (giddiness of the head.) wildness of the eyes, stupor, nausea, vomiting, distention of the stomach and bowels, costiveness, palpitation of the heart, and convulsions.

TREATMENT.--Give an emetic of Indian physic and lobelia, in tincture, these articles may be used either alone or combined. The spirits will stimulate the stomach, and render the operation of the emetic more certain. After copious vomiting has been produced, take common garden rue, and fry it in hogs lard, and give the oil or lard to the patient in table-spoonful doses until the poison is destroyed. Sweet oil is also very good. You should give injections of new milk, with hogs lard or sweet oil in it, until the bowels are well cleansed.

MINERAL POISONS.

SYMPTOMS.--Mineral poisons, when taken into the stomach in too large quantities, soon produces a burning prickling sensation in that part: great pain is experienced in the bowels, accompanied with violent puking, and thirst which cannot be satisfied. It is also attended with dryness and roughness in the mouth and throat as if you had swallowed alum; great restlessness and anxiety. At this stage, unless speedy relief is obtained, inflammation will take place, and soon terminate in mortification, and death will close the painful scene. If the dose of poison taken should not be large enough to destroy life, a fever will ensue, which will last for some time, attended with a constant trembling of the nerves.

TREATMENT.--Give an emetic of American ipecacuanha. Indian physic, or lobelia, or these articles combined: aid the operation of the emetic in every possible manner.--This may be done by applying tobacco leaves steeped in warm vinegar to the stomach. The patient should also take the whites of twelve or fifteen raw eggs, beat well and put into cold water. A gill of this should be taken every few minutes; this will greatly facilitate the operation of the emetic. After the stomach is thought to be measurably relieved of its poisonous contents, give hogs lard or sweet oil, in which has been stewed common garden rue; also give injections of the same in sweet milk.-- Several writers of the old school recommend a puke of white vitriol. I have tried this also, and it had the desired effect.

When there are symptoms of inflammation of the stomach or bowels, refer to those heads for a remedy.

POISONS OF THE SKIN.

Poisons of the skin, such as are often received from poison oak, poison vine, &c., are very painful, and in some instances produce fever. These may be easily relieved by annointing the poisoned parts with night shade and cream. This herb should be bruised, and cream enough added to make an ointment. It may also be relieved by annointing the parts with equal quantities of cedar oil and hogs lard or fresh butter.

MILK SICK.
(Oo-muh-ty-tsu-ni-tlah-e-stee.)

In many parts of the Western country, the inhabitants are subject to this dreadful and often fatal malady. Some suppose that the poison is imparted to the milk by some poisonous vegetable, which was eaten by the cattle.--Others contend, that it is occasioned by the vapors which arise from poisonous minerals in the earth and settle on the vegetables eaten by the cattle. This last opinion is strongly and ably advocated by Dr. Shelton, which I will give in his own words, for the satisfaction of the reader. He says, "This malady is caused by the vapors which arise from poisonous minerals in the earth, and settle on the grass and other vegetables that the cattle eat. This fact is clearly proven by many circumstances. First, by the very appearance of the water, and the rocks, particularly in the lower parts of Indiana, and other sections of country, where it is very prevalent. Second, the very dogs are affected with it from using the water. Third, it seldom makes it attack till in the summer ar fall, after the waters are very low, at which time we know that they contain the greatest proportion of mineral or other impure substances; also, the vegetable substances at this time become tough, and contain much less juice in proportion to the vapour which settles on them. Fourth, if it had been a vegetable which prduced the milk-sick, it would have been found long ago; for, to my own knowledge it has been diligently searched for in many places, by numerous people and not found. Fifth, the scope of land on which it is taken by the cattle, has frequently been ascertained to be very small, and by inclosing it so that they could not get in, the malady was prevented. In this case, if it had been a vegetable of any kind, its growth certainly would have extended over the inclosure in thirty or forty years; for, I am acquainted with a place in East Tennessee of nearly that age. A sixth proof is, that you may let the place remain unenclosed, and the cattle will always be liable to the complaint, as long as there is vegetable matter enough of any kind to induce them to feed on it."--Both people and cattle may have the poison in them for weeks and even months, before they show it: but whenever they are overheated it makes an attack, except on cows which give milk; they seldom die with it, or suffer much from its effects, the poison being carried off in the milk.--People take it from using the milk of cows which are affected with it.

SYMPTOMS.--When the attack comes on, the patient experiences a sense of lassitude, great exhaustion and trembling, from slight exertion. Vertigo or diziness in the head, accompanied with immoderate thirst, burning at the stomach, vomiting, and in most instances, obstinate costiveness. In all cases the breath has a peculiar smell, by which it can be distinguished from any other poison. If the attack should not come on for some time after the poison has been taken, the vomiting is not apt to be so sudden as it is in cases where the poison lays immediate hold on the system.

TREATMENT.--The first object should be to rid the stomach of its contents: for this purpose, give an emetic of the tincture of American Ipecac, or Indian Physic, in table spoonful doses every few minutes, until copious vomiting is produced. The bowels should be relieved of their contents by injections, such as weak soapsuds, in which has been put hogs lard or castor oil, or thin gruel will answer. The ipecac should be continued until the patient is relieved. The vomiting will generally stop when the stomach is thoroughly cleansed of its poisonous contents, and the tincture determine to the bowels: but if this should not be the case, give a little gruel; this will aid in tranquilizing the stomach, and determining the tincture to the bowels. As soon as the stomach has become sufficiently composed. give a mixture of equal quantities of castor oil and spirits of turpentine, in table-spoonful doses every twelve hours. Also give sweating teas, to which may be added a little sulphur. This course should be pursued until the health is restored, which will generally be in a very short time, when compared with the time required by the whites to affect a cure in this disease.

SCURVY.--(Tah-ne-no-loh-quh-tsi-tuh-ne-youh-tso.)

This disease is frequently of a highly putrid nature and generally afflicts persons who have been long confined, without due exercise. Those who have lived a considerable time on salted provisions, or unsound and tainted animal food, or those who have been unable to obtain vegetable food for a considerable time. Bad water, cold moist air, and the influence of depressing passions, such as grief, fear, &c., have a tendency to produce this disease. Neglect of personal cleanliness and debilitating menstrual discharges will produce scurvy in some instances.

SYMPTOMS.--Scurvy may always be known by the softness and spungyness of the gums, which will always bleed from the slightest touch. Ulcers next form around the teeth, and generally eat away the lower edges of the gums, which occasions the teeth to become loose and sometimes to fall out. The face becomes a pale yellow color and sometimes bloated. The breath has an offensive smell. In severe forms of this disease, the above symptoms become greatly aggravated; the heart palpitates or beats rapidly on the slightest exertion; the feet and legs swell, and ulcers break out on different parts of the body, but most frequently on the feet and legs; the urine is high colored; the stools have a very disagreeable smell; pains are felt over the whole body; as the disease advances, blood issues from the nose, lungs, stomach, intestines and uterus--faintings and sometimes mortifications follow.--The appetite remains good to the last, and in many instances there is a perfect retension of the memory until death puts a period to the scene of suffering.

TREATMENT.--It will be extremely difficult to effect a cure in this distressing complaint without the strictest attention to the diet of the patient. All salted animal food must be scrupulosly avoided. If animal food is taken at all, it must be eaten when fresh, but the patient must live chiefly on vegetables; such as scurvy grass, water cresses, garlic, mustard, horse radish, lettuce, &c. may be eaten raw. Cabbage, turnips, parsnips, beets, carrots, &c., may be eaten when prepared in the common manner. The patient's drink should be vineger and water sweetened with sugar; sour butter-milk, lemon juice and water; sour krout is an excellent diet for those afflicted with Scurvy. In the early stages of this disease, and in mild attacks, it may generally be cured by drinking a tea of agrimony, narrow-dock root, sour-dock top or root, or burdoc root. The gums and mouth should be rubbed with the ashes of red corn-cobs twice or three times a day; where the gums are fetid and ulcerated, charcoal finely pulverized, should be mixed with the cob ashes--A dose of the charcoal may be taken once a day if the breath smells disagreeable. The bowels must be kept open by the use of cream of tartar, and occasional charcoal. If the body is affected, the warm or tepid bath, to which add a considerable quantity of vinegar, should be frequently used. Red oak oose, with a little alum in it is a very good wash for the mouth or the ulcres on other parts. Where there is great debility, good wine will be of service; the free use of bitter tonics, will also be found very good. The patient should not neglect to take free but gentle exercise in the open air when the weather is dry and pleasant; but wet and damp weather should be avoided. By attending to the above directions, the disease will be speedily overcome and health restored.

DEAFNESS.--[Tsi-too-ni-leah-no-gah.]

When this complaint is caused by original defect in the structure of the ear, it is incurable. But it is sometimes occasioned by colds, affecting the head by inflammation or bealing in the membrane of the ear, and not unfrequently by the wax becoming hard in the ear.

TREATMENT.--When deafness is occasioned by a cold or inflammation of the ear, take such articles as are reccommended for cold, and steam the ear over bitter herbs; this may be done by putting the herbs in a coffee pot, boiling, them, and placing the ear near the spout: also drop sweet or British oil in the ear. When it is occasioned by hard wax, or by derangement in the auditory nerve, drop a few drops of the tincture of Indian hemlock in the ear, once or twice a day, and about twice a day drop British oil in the ear, about two drops at a time. The smoke of tobacco blown forcibly in the ear, through a quill or pipe stem, will often remove deafness immediately.

EAR ACHE.--(Tsu-ne-le-Squash-te.)

This complaint though painful, often passes off of itself with but very little inconvenience, without resort to medical aid. It often proceeds from colds, inflammation of the internal membrane of the ear, and from insects getting in the ear. This complaint has, in some few instances, produced delirium and convulsions; when supuration takes place, it not unfrequently injures or destroys the hearing.

TREATMENT.--Lard in which onions have been fried, will often give relief, by dropping a drop or two in the ear, and putting a little wool greased with the same in the ear, to exclude the atmosphere or air. If the patient has a cold, he should drink freely of some sweating teas, in order to relieve his cold, by promoting a free perspiration or sweat. In severe cases, drop a few drops of the Hemlock tincture in the ear, or about two drops of the decoction of common tobacco; this should be repeated about twice a day, until relief is obtained. If there is inflamation, the ear should be stoned with herbs as directed for deafness. In ear aches, and in deafness, the ear should be kept stopped with wool, greased with some kind of oil. When ear ache is occasioned by an insect entering the ear drop a few drops of the tincture of camphor, or common spirits in the ear.

HEAD ACHE.--Head ache is often produced by a foul stomach, costiveness, indigestion, or by an obstruction of the circulation of the blood, and not unfrequently it is an attendant symptom of some other disease. But there is a kind of headache which comes on periodically, and is attended with sickness of the stomach, and sometimes vomiting, called sick headache.

TTEATMENT.--When headache is an attendant symptom of some other disease, it will disappear on the removal of the disease which it accompanied. When it is caused by a foul stomach, an emetic will give relief; when produced by costiveness, regulate the bowels by the use of purges. Persons who are subject to paroxysms of sick headache, should live on light diet, take regular exercise, keep the bowels open by the use of cathartics, or bitters composed of equal quantities of gulver and Indian physic, and Moccasin flower root. For a description of these roots, look under their different heads. About the time the fit of paroxysm is expected, the stomach should be cleansed by an emetic of gulver and ipecac or Indian physic; for the mode of preparing and administering this emetic--see Dispensatory. The emetic should be followed by a dose of anti-bilious pills or some other cathartic, if it should not itself operate sufficiently on the bowels. The patient should wear flannel socks on his feet, lined with red pepper constantly.

EPILEPSY--(Epilepsied.)

Persons of all countries, ages and sexes, are in some degree liable to this distressing complaint. The causes which tend to produce it are various and numerous, and the remedies must be varied accordingly. It may be brought on children by worms or by teething; sores suddenly drying up on them, &c. It may be caused by the intemperate use of spirituous liquors, by the sudden suppression of the menses, violent fits of passion, excessive heat or cold, &c.

SYMPTOMS.--Before the fit comes on, the patient is generally troubled with dullness, uneasiness, giddiness, pain in the head, palpitations of the heart, disturbed sleep, and difficulty of respiration. The complexion becomes pale, and the extremities cold. Females, it is said, are most liable to this disease. It is sometimes caused by some natural defect of the obstructions of the blood vessels.

TREATMENT.--When a person is seized with a fit, a piece of wood or a spoon should be placed in the mouth, to prevent the tongue from being injured by the teeth. When an obstruction of the brain is feared, bleed in the foot, and evacuate the bowels as speedily as possible, by the use of laxative injections. If worms are the cause expel them as directed under that head; if teething, bathe the feet in warm water frequently, and apply plasters of mustard seed to the feet, to produce a revolution from the head, and at the same time keep the bowels open by cooling injections; if customary evacuations have been stopped, they should be restored; if indigestible food or spirituous liquors taken into the stomach is the cause, give an emetic; if weakness and irritability of the nervous system is the cause, give the atmospheric tincture. In all cases of this disease the powdered root of the moccasin-flower should be taken freely--dose, a tea-spoonful of the powdered root in a pint of boiling water. To prevent a return of the fit, keep the bowels open, take a chalybeat pill morning and night, and drink a tea made by putting a tea-spoonful of powdered mistletoe (taken from the white-oak tree) and a tea-spoonful of the powdered root of the moccasin flower into a pint of boiling water. All possible pains should be taken to keep the mind at ease and cheerful, and to prevent the intrusion of violent and agitating passions.--When Epilepsy proceeds from natural defects it is incurable.

APOPLEXY.--(Apoplexia.)

Apoplexy is a sudden deprivation of sense and motion, while the heart and lungs still continue in regular action.

CAUSES.--Intense study, violent passions, wearing the neck-cloth too tight, luxurious diet, suppression of urine or other discharges, sudden checks of perspiration, hard drinking, excess of venery, too large doses of opium: in short, whatever determines or throws so great a quantity of blood to the brain that it cannot return from that organ, has a direct tendency to produce this distressing and often fatal disease. Persons who lead an inactive life--persons of advanced age, corpulent habit, short neck, and large head: also, such as live on full, rich diet, are more liable than those of the opposite habits.

SYMPTOMS.--It is usually preceded by giddiness and swimming of the head, loss of memory, night-mare, noise in the ears, drowsiness and difficulty of breathing. It sometimes though rarely comes on suddenly and cannot be accounted for--and as it goes off leaves some part of the system in a *paralyzed* condition, which is then called Palsy. (For treatment of such cases see Palsy.)

TREATMENT.--Raise the patients head, place him where he can breathe cool air, and remove every thing from about the neck, that has a tendency to compress it. If the patient is robust and of a plethoric habit, bleed copiously in the foot, bathe the feet in warm water, and then apply plasters to the feet in order to produce a revulsion from the head. Apply cloths wet in cold vinegar and water to the head, changing them as fast as they become warm. Evacuate the contents of the bowels by means of a purge, aided by injections, a portion of each injection should consist of anti-spasmodic tincture, or some of the articles described in the class of anti-spasmodics in Materia Medica. If the patient should be old and feeble and the countenance palid, he should be used sparingly, the head should be raised and frequently turned; if the patient can swallow, give a purge and aid its operation by injection, use the anti-spasmodic tincture freely by the mouth and by injection; rub the feet with the anti-spasmodic tincture and apply plasters of mustard-seeds, wet with the same. After the violence of the fit has subsided, follow the same course as directed for *Epilepsy*.

Persons afflicted with either of these dangerous diseases, should live on spare diet, and carefully avoid all predisposing causes.

VENERIAL.--(Tsu-ne-nu-sup-huh-skah.)

The prevalence of this *filthy* disease among mankind, is another proof among the many that might be adduced, that it is the interest of mankind to be virtuous, if they would be happy, and he that would be healthy must be temperate. At what time and place this disease had its origin, is now unknown to the medical world; but it first attracted attention in Europe, about the close of the fifteenth century, and was communicated with great rapidity to every part of the known world, and became such a desolating scourge to the human family, as to render it an object of great medical attention.

This complaint is produced, in most cases, by a healthy person having sexual intercourse or connexion with another who has this infectious disorder in the genitals or privates; and most frequently occurs among persons of illicit habits, and hence disgrace is attached to it: and on this account, may have been induced to conceal their situation, until, by endeavoring to hide their shame they have ruined their constitutions. Yet it sometimes happens, that this contagious complaint, is caught innocently; but the difficulty of proving innocence, almost always leaves a blight upon the character of the sufferer. After this disorder has been taken in the manner I have described, it will depend very much on the state of the system and other peculiarities of the system not distinctly known, at what particular time the disease will make its appearance. In some persons, whose systems are very irritable, it will show itself on the third or fourth day after sexual connexion with a person infected with the disorder; in other persons, it will be eight or ten days, and even a longer time before it makes its appearance. In fact, cases are mentioned by good medical writers, in which the venerial matter has remained as it were asleep in the system, for a much greater length of time. Some say one month, some three, some six, and others a year and so on. But I suspect the fact to be in those cases in which the disease is supposed to appear after a considerable time, that the persons have not been entirely cured; or, in other words, that the disease has merely been driven back by quackery, and afterwards showed itself under the following forms: In the throat, in the eyes, in the nose, on the legs, in swellings of the groins, in splotches or sores on the body, &c. When it makes its appearance in the above forms, it is called *Constitutional,* because it is firmly seated in the whole body by the venerial poison having been absorbed and carried into the whole circulation. The venerial disease is very contagious. I copy the following from the writings of Dr. Gunn: "The venerial disease may be communicated by wounding or pricking any part of the body with a lancet, having on its point any particle of this venerial poison. I recollect a student of medicine, who came very near death from cutting his finger slightly when dissecting a person who had died of the venerial disease; the poison matter was communicated to the slight cut in twelve hours afterwards; he labored under violent fever, which continued ten or twelve days, before the inflammation could be subdued. This disease may also take place from an application of the matter to a scratch, to a common sore, or to a wound. Several instances are mentioned of venerial or pox sores being formed in the nostrils, eye-lids and lips, from the slight circumstance of persons having the disease touching their nostrils, eyes, or lips with their fingers, immediately after handling the venerial sores on their own privates. These remarks are made with the intention of showing how easily this loathsome disease, with all its impure and life corrupting taints may be communicated, and to place physicians and individuals on their guard against infection."

Venerial disease have two distinct forms. The first is Pox, properly so called, and the second, Clap, called by
physicians *Gonarrhoe*. There is also another form, which however, always arises from one of the other two, or from both in combination, and is nothing more than the one I have before described as constitutional.

POX.--[Oo-ni-lech,]

The Pox is a most contagious, corrupting, dangerous and destructive disease, and if suffered to progress in its ravages on the human body, never fails in desolating the human constitution, or destroying life at its very core.--It has two forms. First, *local,* and second *constitutional,* When it is first received by cohabition, it is for a while located in and confined to the *privates* and *genital* organs; but, if let run on for a length of time, without being cured, it affects the whole system, and deranges and impairs the constitution--it is then called *constitutional.* It is very readily communicated by sexual connexion, or from either the father or mother to the offspring; also from the mother or nurse, who gives suck to the child; and we also have it from very respectable authority, that it may be taken from inhaling the breath of a person who is affected with it, or by kissing, drinking, &c., with such persons, or by washing the clothes of those infected with it, &c. Pox differs very much from *Clap,* in the length of time in which it makes its appearance from the time of its reception. It will sometimes make its appearance in seven or ten days, and sometimes it will be two or three weeks, and even longer before it breaks out.

SYMPTOMS.--This disease generally makes its appearance by what physicians call *chancres,* and when taken from sexual connexion with an infected person; the first warning of its approach is generally an itching about the head of the *penis,* or on the side of the penis near the end, and on the inside of the lips of the *privates* of females.--Little pimples soon rise and fill on the top with a whitish or yellowish looking matter; in a very few days, these pimples enlarge themselves and become what are called venerial sores or ulcres. These sores sometimes, after a long continuance, gradually disappear and others break out at the same time. Sometimes the first sores continue to enlarge as a kind of eating ulcer, with hard looking edges, and discharge a thin unhealthy matter. The Pox also, sometimes makes its appearance, in what are called

buboes. These are hard lumps like kernels or swellings which rise in one or both groins. These swellings gradually increase in size, until they become about the size of an egg and have an angry red color, and unless driven away by the application of medicine, they will come to a head and discharge a thin briny looking matter. These *buboes* generally produce great pain and some fever. *Buboes* sometimes make their appearance in the arm-pits, in the throat and about the neck. These last appearances of *bubo* however, are not very frequent, and are often the effects of mercury improperly administered in the Pox arising from the disease itself. When the constitution is very irritable, the disease will sometimes attack the nose, the throat, the tongue, the eyes, the shin bones and so on. In some cases the whole of the nose and palate bones have been eaten out, and the nose flattened down almost to the upper lip. When the disease has been communicated from parents to the offspring, it sometimes comes into the world full of sores, and sometimes skinned and raw nearly all over. How the feelings of virtue and common decency must recoil at such a disgraceful and yet heart-rending and truly pitiable sight.

TREATMENT.--First give an emetic of gulver and wild ipecacuanha, or Indian physic, prepared as follows: Take two ounces of gulver root, and one ounce of ipecac, or Indian physic, (the root) put them in one gallon of water, and boil down to a half pint, and give this in half gill doses at intervals of fifteen minutes, until vomiting is produced. As soon as this medicine commences operating, give warm water to assist its operation: when the stomach is sufficiently cleansed, give water gruel to determine it to the bowels. You may also give a teaspoonful of the flour of sulphur with the gruel, to remove the sickness from the stomach. After the operation of the emetic, the patient should drink of the following decoction: Take a handful of white sarsaparilla root, the same quantity of yellow sarsaparilla root, a double handful of wild mercury root; boil all these articles together in three gallons of water down to a half gallon. The patient should drink of this decoction three half pints a day. Also give powders, made as follows: Take a double handful of agrimony, the same quantity of bamboo brier, two ounces of Indian hemp; wash these roots clean and dry them in the shade, where they will not be exposed to the dampness of the rain or dew; when perfectly dry, pulverize and bottle up for use. Of these powders, the patient should take a dose, a teaspoonful; if this should fail to operate in two hours, give half a teaspoonful, and repeat every two hours until it purges. After the first dose, which should purge well, give a sufficient quantity each day to keep the bowels regulated. If the above powders should fail to operate on the bowels with sufficient activity, give a dose of calomel, say from twelve to twenty grains, according to the constitution of the patient; after which give a moderate dose, say five grains twice a day for two days; the third day give one dose, after which give three grains a day, until the gums and root of the tongue feels slightly sore. On the first appearance of the soreness of the gums and root of the tongue, the use of the calomel must be discontinued, and a tablespoonful of the flour of sulphur taken twice a day; also castor oil should be taken twice a day; a tablespoonful each time.

DIETS.--When the patient is of full habit of body, the diets should be light, such as light soups, buttermilk, bread and shortening, a little chicken or squirrel cooked in its own oil alone, or something of the kind. But if the patient is in delicate health, or much reduced by the disease, or by the use of strong medicines, nourishing diet, wine and tonics will be proper.

CLAP.

This disease may be communicated by sexual intercourse. Or if a woman be afflicted with it while pregnant, unless she be cured before the birth of the child, it is sure to have it; some children are born with it, whose mothers have been cured before delivery. These are the only means by which this disorder can be communicated. When taken by *sexual connexion* in sound persons who have never had it before, it will not in general show itself sooner than from five or six to nine days, but I have never known it to go over the tenth day after it was received, before it made its appearance. Persons who have frequently been afflicted with Clap, or those of weakly, irritable habits, are apt to discover it on them about the third day, and it very seldom passes the fifth with such persons, without showing some signs of its approach.-- The mode of life will make a difference in the length of time in its making its attack. If you are temperate, it will not show itself so soon as if you are intemperate after the reception of the disease.

SYMPTOMS.--In most instances, the first symptoms are an itching, and slightly painful sensation about half an inch up the water passage from the end of the penis, and burning or scalding sensations or feelings in the urethra, or canal of the penis, whenever you urinate or make water. In a short time, say in the space of from five to twenty hours after these symptoms are felt, there will be a slight discharge of matter from the privates, nearly like the white of an egg; in a short time it becomes more copious, and of a yellow color, and lastly, of a greenish color. In a few days the soreness extends up the water passage (urethra) to the neck of the bladder, and the system is thrown into a general feverish condition. In females this disease is somewhat more simple than in males. In its first stage it resembles the whites in their worst stage, and they can go much longer, and suffer less without a remedy, than a male; because the parts are larger, and the matter more freely discharged, before it becomes so irritating. But the female labors under this disadvantage: The disease sooner passes up, both the birth place and water canal, and affects the womb and bladder both. Sometimes the tisticles of the male swell, and become very painful, and the penis inclines to great erections, thereby giving great pain. The Clap shows itself much sooner, and progresses much more rapidly, and soon becomes more violent on those who have it frequently, or more than once, than it does the first time; in such cases it frequently attacks the whole water passage at once, or perhaps up near the first place. After the disease has been suffered to run on for a length of time, the eyes become weak, and the little veins are all engorged with blood; the edges of the lids look swelled, inflamed and hard, the hollow around the eyes assume a dark appearance, and the countenance wears a *dull, dejected,* and *sickly* aspect.

TREATMENT.--First cleans the bowels with anti-billious pills or some other cathartic; and then drink freely of a decoction made as follows: Take of white sumac root five pounds, of the small kind red root two pounds, one double handful of black or dewberry brier root, a double handful of persimmon bark, (the bark of the root is preferred) boil all these articles in ten or twelve gallons of water, down to a half gallon, and strain for use. Of this decoction the patient must drink freely, and keep the bowels regulated by the use of anti-billious pills or some other cathartics. He should also take a handful of sarsaparilla, and boil it in two gallons of water down to a quart, and take of this decoction a half pint or more each day. The above medicines should be used perseveringly until the discharge ceases. The patient should eat no strong diet; a little squirrel or chicken may be taken if cooked with but very little seasoning, if wheat bread is used it should not be shortened. Tea or coffee may be drank, but ardent spirits must be avoided or the remedies may be discontinued.--The patient should avoid all exposure or violent exercise, and all sexual intercourse with others.

ANOTHER MODE OF TREATMENT.--Cleanse the bowels as above directed, and then drink freely of a strong decoction of equal quantities of dewberry brier root and blue flag or gleet root.

In all obstinate cases where the above remedies fail, give a dose of calomel, say from ten to twenty grains, and work it off with castor oil or rheubarb, and then drink freely of one of the above decoctions, in coonexion with the sarsaparilla decoction.

There are many other preparations which may be used to great advantage in this filthy complaint, as may be seen in the Dispensatory. But those afflicted with this loathsome disease, are too apt to be changing the mode of treatment before it has had a fair trial. This is very wrong, and should by all means be avoided if a speedy recovery is desired. After having pursued one remedy for ten days without an alteration for the better, it may be changed for some other mode, but not sooner, and temperance must be strictly observed.

BLOODY URINE.

A discharge of Bloody Urine may be occasioned by the lodging of a small stone in the ureter, or in the kidney, which wounds the part with which it comes in contact; when in this way it usually deposites a sediment of a dark brown color, sometimes clotted, and is attended with an acute pain and sense of weight in the back, and difficulty

in making water; when the above symptoms occur, and it is supposed to be caused by the lodgment of a stone, look for a remedy under the head of Stone and Gravel.--When the blood proceeds immediately from the bladder, its discharge is usually accompanied with a sense of heat, and pain in the lower part of the belly.

TREATMENT.--Give a tea of the powdered root of Jerusalem oak, a table-spoonful to a pint of boiling water; also give powders of egg-shells and alum, equal quantities.--The egg-shells must first be browned or parched in an oven and reduced to a fine powder; they should be finely pulverized, and the two articles mixed together. Dose, a tea spoonful night and morning. The bowels must be kept open by the use of anti-bilious pills or some other cathartic, such as rheubarb, oil, &c. Bitters of yellow sarsaparilla in common spirits will be found of great advantage in this disease.

GRAVEL AND STONE.
[Tsu-ni-nic-luh-huh-sko-oh-tekoh-luh.]

Gravel and Stone, though distinguishable from each other, appear to originate in the same causes, and require similar treatment. Gravel is usually understood to mean *calculi,* (from the old word *calx*) a *limestone,* or *little sand-like stones,* which pass from the kidneys through the ureters into the bladder. Stone is a strong concretion of matter, which enlarges, and hardens by time; seldom found in the ureters or tubes themselves, but generally lodged in the kidneys or bladder; when the Stone is in the kidney it is because it is too large to be passed off through the ureters (ureters are small tubes which extend from the kidneys to the bladder and convey the urine into the latter) into the bladder; and when found in the bladder, it is from the fact of its being too large to be passed off through the channel of the penis.

SYMPTOMS.--When a disposition to Gravel exists in the urinary system, there will be occasional paroxysms or fits of pain in the back, which sometimes shoot downward to the thighs, and sometimes a numbness of one of the legs inside, accompanied with a retraction or drawing up of one of the testicles or stones in men, almost constant desire to make water which is attended with the most agonizing pain; and is sometimes terminated by a discharge of small gravel stones from the urethra with the urine, sometimes. The stone which is usually found in the *kidneys* or *bladder,* sometimes in both, is a disease of more serious and dangerous consequences. When the *stone* has acquired some size, if situated in the bladder, there is an almost constant desire to make water, which is voided in very small quantities, sometimes, drop by drop with great pain--and sometimes in a small stream, which occasionally stops short, and is attended with almost insupportable pain. In some persons, the violence of straining to void the urine, makes the rectum or lower gut contract, and expel its excrements; or if that gut be empty, this straining occasions *tenesmus* or a constant desire to go to stool. There is often blood to be seen in the urine, and sometimes pure blood is passed off in small quantities.--When calculus or stone is formed in the kidney, in addition to the general symptoms of stone in the bladder, there will be felt a dead, heavy, dull pain in the loin, where the kidney containing the stone is seated, frequently accompanied with chillness or creeping coldness, in and over the part affected. In severe cases of calculous or stone, either in the kidneys or bladder, there is frequently experienced, during the time of passing the urine, sickness of the stomach, a desire to vomit and much faintness. The gravel and sometimes stone, when the latter has not become too much enlarged from the lapse of time, may much more easily be removed from the bladders of females than from males.

TREATMENT.--The bowels should be kept open by mild and cooling purges, such as cream of tartar, American senna or some similar article. There are two small bones in each drum fish's head, they are nearly round, and about the size of the thumb nail, reduce a number of these bones to a very fine powder, and take a teaspoonful of these powders morning and night. This is a certain remedy for stone or gravel, it will dissolve the stone, and cause it to be passed off in the urine.

There are many other valuable articles for this truly painful and distressing complaint. A decoction of cat tongue, drank freely seldom failes to give relief. The Horse radish is very good, and the common garden radish is also good; the proper mode of using the common radish is in decoction, or the expressed juice; this is a valuable remedy, and it is said by many to possess the property of dissolving and carrying off the urine.

The index will point to numerous valuable articles for suppressed urine. Every thing of a heating nature, both in diet and drink, must be strictly avoided.

The patient should take regular, but moderate exercise. The index will also refer the reader to several valuable preparations for gravel in the dispensatory.

Turk in his pamphlet, which was published in 1843, says, "I have cured this disease of long standing, in four or five days, with a root in alcohol or good spirits; I have never failed, and until I do, I will continue to believe in its efficacy. I could cite many cases of the most notorious kind, but as I intend a small work, I will give but one. A gentleman in North Carolina had been afflicted for 15 years, and so much so, and so many trials had been made, that all hopes of a cure was abandoned; he had to draw his water with a catheter; he called upon me, and I furnished him the medicine; he declared to me that I cured him in three or four days; I saw him occasionally for three or four months, and he remained well." The root I cure this disease with, grows in every part of the United States, where I travel; it grows on upland, and low grounds--I have mostly found it in fence corners; in good land, it grows from five to eight feet high, has a bushy top and yellow blossoms, and blooms from the last of July, and through August. The root is the part for use, and hard to dig--has many roots growing horizontally, when dug out, and exposed to the sun, inclines to turn red, between the bark and the woody part of the root is a sticky rosin. The rosin contains the medicine, the stalk while green has the appearance of four square, but when close examined, it is found to be round, and four feather edges. As I know of no other root that contains this kind of rosin, I cannot see how it can be mistaken, and another proof, it makes a very pleasant drink, and will remind us of the pure rosin. This medicine acts directly upon the water passage, and will dissolve any stone in the kidneys or bladder. It seems to have nothing to do with any other part of the system; the gravel, after being dissolved, is thrown off, and looks like lime. The mode of preparing and using it is as follows. After the root is dug and well

washed, cut fine and fill a quart bottle with the root, then put in the spirits; and it will soon be fit for use. The bottle will bear filling two or three times, without renewing the roots. This tincture should be drank as freely as possible, according to the age, strength, and constitution of the patient,--it should be used most freely from 10 to 4 o'clock in the night. In all diseases, the system should be in good order, and costiveness in this disease should be particularly avoided. This root will cure any inflammation of the bladder, diabetes, or involuntary flow of urine. I have never tried the above remedy, as the pamphlet very lately came to my hands, but thought proper to give it a place, as the root abounds in our country, and it certainly deserves a trial.

The Uva Ursi of the mountainous regions of Europe, and possibly of this country, is said by physicians who profess to be acquainted with its medical virtues, to stand unrivaled as a remedy for gravel or stone. A full description of this article may be seen under its proper head to which is added several cases of actual experiment, and the result in cases of gravel or stone.

DIABETES.

Which means an immoderate flow of urine, commonly without any particular pain in the urinary organs.

SYMPTOMS.--The quantity of water usually discharged in this disease is more than double the quantity of liquid taken both in food and drink, and it is as transparent as spring water--it has a sweetish taste like sugar and water, and a very faint smell, as if mixed with rosemary leaves. After this disease runs on for a length of time, the mind becomes dull and melancholy, the skin dry and hot, immoderate thirst which cannot be satisfied, the appetite becomes voracious. There is a gradual emaciation of the whole body, attended with great debility, a sense of weariness and great aversion to motion. There are frequent darting pains in the privates, accompanied with a dull, heavy pain in the small of the back, the bowels are costive, the pulse irregular. As the disease advances, fever takes place, similar to that in hectic and consumptive cases, the feet begin to swell, and death in a short time ensues. The attacks of diabetes are generally slow and gradual. It is sometimes two or more years in making its advances on the constitution.

TREATMENT.--Cleanse the stomach with an emetic, follow the emetic with a cooling purge. Purges should not be used in this disease any further than to prevent costiveness. The patient should drink daily of a tea of piny-weed or gleet-root or both combined, and take a chalybeate pill night and morning.

The diet should be nourishing, principally flesh; vegetables should be avoided.

CHOLIC.--[Tsu-ne-yoh-low-tis-scoh.]

Of this disease there are generally reckoned three kinds, as flatulent, bilious, and nervous or cramp Cholic. The causes which predispose to an attack of Cholic, are flatulence, indigestible food, unripe fruits, fermenting drinks, windy vegetables, excess of bile, costiveness, sudden check of perspiration, cold, worms, other diseases improperly or imperfectly cured, &c.

SYMPTOMS OF FLATULENT CHOLIC.--This Cholic may be distinguished by a rumbling in the bowels, and a disposition to belch or discharge wind from the stomach. It is also attended with sickness and sometimes vomiting, a violent pain is felt in the stomach, and if undigested food has passed from the stomach into the bowels, the pain will extend from the stomach to the bowels and be most severe at or near the naval. The stomach and bowels both being distended or tightly swelled.

TREATMENT.--This kind of Cholic may generally be relieved by very simple means, such as a drink of warm toddy, to which may be added fifteen or twenty drops of oil of penny-royal, essence of peppermint, or strong mint tea--a tea of black or red pepper, ginger, spice, calimus, dog wood berries, are all good and will generally give relief; a tea of bubby-root is also good. Hogs hoofs burnt and reduced to powders, taken in tea-spoonful doses mixed with honey, every few minutes, will be found an immediate remedy. Spikenard-root taken in powders or bitters, is very good; spirits into which has been put balm of gilead buds, is excellent, a dram of camphor is good. Persons who are addicted to Cholic should use balm of gilead buds, spikenard-root, prickly-ash bark of the root,

or asafoedita digested in common spirits for a daily bitter. These articles may be used alone or combined at pleasure. In a violent and stubborn attack, bleed and bathe the patient; if bathing vessels are not at hand, bathe the feet, and apply cloths wrung out of hot water, to the belly, as hot as can be borne, warm salt, applied in the same manner is very good. If costiveness prevails, give laxative clysters, such as butter-nut syrup and gulver tea combined; the clysters should be continued until the contents of the bowels be thoroughly evacuated, after which give a few clysters of new milk and water.

BILIOUS CHOLIC.

SYMPTOMS.--Loss of appetite, a bitter taste in the mouth. an acute pain about the naval, costiveness, attended with sickness and vomiting of a bilious matter.

TREATMENT.--Open the bowels with some active cathartic, aided by injections. If the pulse be frequent and high, bleed. After the bowels have been thoroughly cleansed, give a few injections, composed of new milk and water, with a little hogs lard in it. When the stomach is irritable, it may be quieted with mint tea, or a tea of cholera morbus root; peppermint bruised and applied to the pit of the stomach, will aid in checking the vomiting. After relief is obtained, it will be advisable to take an antidyspeptic or hepatic pill night and morning for a few days. This will restore the liver to healthy action, and increase the tone and strength of the stomach.

NERVOUS CHOLIC.

Nervous or Cramp Cholic may be distinguished by a disposition to cramp, accompanied with pain similar to other Cholic.

TREATMENT:--Bathe the feet in warm water and bleed in the foot. If the blood can be taken from the vein which lies nearest the ankle-bone on the inside, it is preferred. Give laxative injections--butter-nut and gulver if it can be had. At the same time, let the patient drink freely of sampson snake-root tea, a tea of bubby-root, or the root or bark of the prickly ash. Warm applications to the pained part, such as cloths wet in hot water or warm salt, will aid in subduing the spasm or cramp Cholic, sometimes it extends to the blood; this kind of Cholic produces a dull, heavy sensation throughout the whole system; this cholic is most common among pregnant women, and may be relieved by keeping the bowels open, by the use of mild cathartics and drinking freely of bitters composed of spikenard, white sarsaparilla and the bark of the root of prickly ash. These articles may be used all together, or either of the articles used alone. The expressed juice of prickly ash root is not surpassed by any other remedy in our knowledge for Flatulent or Nervous Cholic.

LOCKED JAW.

Locked-Jaw may be considered an involuntary cramp or contraction of all the muscles of the body. It most frequently arises from wounds, and in some instances from very slight wounds, such as the slight puncture of a pin, needle, or splinter under the nail; also from cuts, snags, &c.

SYMPTOMS--Are a dull stiffness of the head and neck, an uneasiness in the breast, soon followed by a change of speech, and a difficulty of swallowing, the patient frequently remains perfectly in his senses. The stiffness in the jaws gradually increases, until the teeth becomes clenched, the tougue also becomes stiff, and violent and alarming proxysms now take place in the muscles.

TREATMENT.--When this painful disease proceeds from a wound of any kind, the wound should be immediately opened and cleansed of any matter that may be in it; then fill it with spirits of turpentine or warm salt, and cover it with a warm moist poultice. If the patient can swallow, give a purge of gulver and butter-nut syrup combined, or caster oil; the cathartic should be aided by injections.--For preparing and administering injections, see under the head Clystering. Also give a tea of seneka or black snakeroot. If the patient cannot swallow, give injections freely of some active cathartic, as it is all important to have the contents of the bowels immediately evacuated. Tobacco or tincture of lobelia may be combined with the purgative injections to great advantage. The warm bath should never be dispensed with in this alarming disease, as it will aid both in relaxing the muscles and in the operation of the cathartics, whether administered by injections or otherwise. For an external application to the jaws, use the red pepper poultice, a thin cloth should be put between the skin and the poultice to keep it from burning the skin. The poultice should be wet with antispasmodic tincture, and from a half to a whole tea spoonful of this tincture should be taken internally, repeated at intervals of ten or fifteen minutes or oftener, as circumstances require it. If the patient cannot swallow, this tincture should be put into the mouth, it will soon find its way to the root of the tongue and will aid greatly in relieving the spasm and cause the parts to become relaxed. A poultice of life-everlasting or sassafras buds is very good, applied to the jaws and throat, to produce a relaxation of the parts.

TOOTH-ACHE.--(Oo-yoh-quah-li-skee.)

This excrutiating and distressing complaint, is thought by many persons to originate in the teeth. This idea, however common it may be, is very erroneous. It is in most instances, a symptom of other diseases, which diseases must be sought out and removed before relief can be obtained.

TREATMENT.--When this disease is supposed to arise from rheumatism, look under that head for a remedy.--When it is caused by the stoppage of periodical evacuations in females, refer to that head among the diseases of women for a remedy. When it is attended with costiveness, headache and fever, purge freely with salts, caster oil, rheubarb or antibilious pills. Toothache is frequently caused by the nerve of the tooth being exposed to the atmosphere. When this is the case, wet cotton or lint in ceder oil, essence of peppermint, penny-royal or spirits of camphor, and plug the hollow of the tooth with it, renew this lint frequently and it will generally give relief; cotton or lint wet with laudanum is also very good. Toothache sometimes arises from rheumatism, when this is the case, both the sound and decayed teeth will be pained; there will also be a dull heavy pain extending along the jawbone, and the whole side of the face will be affected to a greater or less extent. When this is the case, wilt the leaves of the Jamestown-weed (jimson) by putting them in hot water and then pressing them tightly in the hand, then put them on a cloth large enough to cover the pained part, and bind them to the jaw as warm as can be borne: when the leaves become cold, warm them by pouring a hot tea of the same leaves over the plaster, and again apply it. This will in most cases, give relief if properly attended to. Extracting should be the last means resorted to for relief. I have known sound teeth extracted, and afterwards other diseases were discovered to be the cause, which had to be removed before health could be enjoyed. Toothache is very common among pregnant women, particularly during the first stages of pregnancy; cold is almost always the exciting cause; it may be relieved by the use of mild cathartics, bathing the feet in warm water and drinking some diaphoretic teas.

BEALED JAW.
(Coh-you-cah-tsi-tuh-nu-tis-lay.)

Bealed Jaw is generally caused by cold settling on a decayed tooth. The patient should drink freely of some sweating tea, such as pennyroyal, balm, mint, sage, &c. Also steam the wound over bitter herbs, such as catnip, hoarhound, &c. For an external application, I have never found anything so good as the Jamestown leaves, (jimson) as directed above for Toothache. This, if applied in any reasonable time, will allay the inflammation and prevent it from bealing. The leaves may be procured in summer, and cured in the shade and made into a poultice when needed. In this way they may be had at all times in the year.

PALSY--(Tsu-ni-luh-tah.luh-uh-skah.)

This disease is characterized by a numbness or want of feeling in the part affected. It sometimes affects one part and sometimes another. If it attacks the heart or lungs it must inevitably prove fatal--its danger or fatality is to be expected in proportion to the vitality of the part affected.

Causes which predispose to an attack of Palsy, are apoplexy, obstructions of necessary evacuations, excessive venery, any injuries that obstruct the passage of the nervous fluid to the brain, to the organs of motion, injuries of the spinal marrow, intense study and great distress or anxiety of the mind. In short, any thing that has a tendency to weaken and relax the nervous system in an extreme degree, has a tendency to produce this alarming disease.

SYMPTOMS--Are giddiness, torpor, uneasiness in the head, failure of the memory, dullness of intellect, coldness, tremor, creeping and pain in the part affected.

TREATMENT.--First dig a pit in the ground, just deep enough for the patient's shoulder to be even with the surface of the earth when sitting on a chair in the pit; build a fire in the pit and let it remain until the pit becomes hot, then take the fire out as soon as the pit becomes cool enough for the patient to bear it; place him or her in it on a chair, and cover it over with a blanket, only leaving the head of the patient to the air. Let the patient remain in this pit until copious sweating is produced, or as long as the constitution and strength will allow. When the patient is taken from the pit, scarify the joints of the affected part, and very especially the back bone, and wash the scarified parts with the tincture of Indian hemlock. For a full description of this shrub or bark refer to that head. The mode of preparing the tincture is as follows: Take of the powdered leaves of Indian hemlock a table spoonful, put it into a half pint of hot water, and let it remain thirty minutes or more, then wash the scarified parts with this infusion or tincture, every hour for twelve hours, if the natural feeling does not return in a shorter period; but if it should return, (which is not uncommon under this treatment,) the use of the hemlock should be discontinued, and the parts should be frequently and well bathed with strong vinegar. The patient should drink freely of a beer made as follows: Take malt enough to make ten gallons of beer, put it into ten gallons of water, add to it one quart of finely ground mustard seed, let this beer ferment (work or become sour) and it will be fit for use. The patient should drink freely of this beer until health is restored. Diets should be light and nourishing, and if the bowels be costive, they should be relieved by mild purges, and injections.

WEN.

Wen is a fleshy substance growing between the skin and the natural flesh, without any known cause. When it makes its appearance on the neck it frequently grows to such an enormous size, as to render breathing very difficult.

TREATMENT.--First annoint the Wen with cedar oil.-- (For the mode of preparing this ointment, look under its proper head.)--It should be annointed with this ointment every twelve hours, then apply a plaster made of equal quantities of common soap, and table salt, mixed well together; the fourth day this poultice should be changed, for one made by boiling the pulverized root of maycock in water, and thickening it with corn meal. If the leaves are green, they may be applied if preferred, being first bruised or wilted. These poultices should be changed occasionally, keeping the first kind on about one third of the time, and the maycock the other two-thirds.--The wen should be regularly annointed with cedar oil every twelve hours; the ointment should be bathed in with a warm iron.

Dr. Wright gives the following easy and simple remedy as an infallible one: "Wash it with common salt dissolved in water, every day, or make a strong brine of alum salt; simmer it over the fire; when thus prepared, wet a piece of cotton in it every day, and apply it constantly for one month, and the protuberance will disappear." I have never tried Dr. Wright's prescription, but have no reason to doubt its efficacy. It is at least so simple, and so easily tried, that I have thought proper to give it a place.

DISEASES OF THE SKIN.

The close connexion which exists between the stomach, bowels, and skin, is evidently demonstrated by the fact, that in many instances where the stomach and bowels are internally disordered the skin exhibits external evidence of the disease. Many eruptions which show themselves on the skin, are positive proofs of the deranged state of the system internally. Care should therefore be taken to ascertain the true cause of those eruptions. If they are produced by no impure state of the blood, foul stomach, costiveness, or from some constitutional disease derived from parents; if either of these causes produce eruptions of the skin, it will be obvious to all, that it is to be removed by internal remedies. When eruptions are produced by the above causes, and should be driven from the skin, by external remedies alone, it will generally produce, and in many instances, seat some fatal disease on the vital organs. Dr. Gunn says: "Whenever diseases exhibit their effects on the skin, you may be assured that they are the

efforts of nature to relieve herself from oppression; and, the real business of the physician is, to assist nature, and never to retard or stifle her operations." The cause of eruptions should be sought out and removed, and when they proceed from internal disease, they will have to be removed by internal remedies.

TREATMENT.--The first thing to be attended to in case of eruptions, is the bowels, which must be cleansed by a cathartic, and then kept open by the use of salts, cream of tartar, and sulphur, or a free use of yellow poplar root bark, taken in decoction, bitters, or powders, as may be the choice of the patient. A tea of sarsaparilla, sassafras, or spice wood drank cool, will be found beneficial. For external applications use common starch or flour. These simple applications will be found a cooling, and pleasant remedy: they will allay the itching and uneasiness of the skin for a time, and when these unpleasant sensations return, the application should be repeated. Persons that are subject to eruptions of the skin, should live on light and refreshing diet, and avoid everything of a heating nature either in food or drink. They should also keep their bowels open by the use of cooling cathartics as above directed. The skin should be kept clean, and moist, by frequent bathing, and washed in warm water.

SCALD-HEAD.
(Tsa-nah-li-stah-wo.)

Scald-head is an inflammatory eruption of the skin of the head. It generally commences with an uneasy tingling, itching sensation, as though something was crawling through the hair. In a short time, numerous small white pimples, arise at the roots of the hairs, under which are very small ulcers, which will in a short time, discharge a whitish matter. At other times it commences more boldly, and presents clusters of small red pimples or pustules, dispersed throughout the head. Some even advance to superation, leaving pits filled with pus; many scabs fall off like bran, while others adhere closely to the skin. Children are most liable to this disease, but the adult is not exempt from it. Neglect of cleanliness, bad nursing, or a want of a due portion of wholesome food may produce this disease. It is contagious, and is often taken by children from wearing the hat or cap of a person affected

with it, by sleeping in the same bed, or by combing with the same comb. It also descends in some instances, as a hereditary disease, and when it occurs in connexion with a scrofulous habit, it is tedious and difficult to cure.

TREATMENT.--The hair must be shaved or trimmed, as close as convenient, and the head well cleansed with warm water and soap; the first attempt to remove the scabs, and cleanse the head, will require some time, in order to soften the scabs that they may be removed without pain. When this is done, the head should be washed in a decoction of the small kind of smart weed; (for a description of this weed, look under that head;) after the head has been washed in the above decoction, take a feather and dip it in ceder oil, and touch it lightly on the deepest pustules or scalds. Then anoint the head with an ointment made by stewing the bruised smart weed in hog's lard or fresh butter. Next apply a poultice of corn meal mush, made without salt; after it is spread on a cloth large enough to cover the affected part, sprinkle on its surface a thin coat of the flour of sulphur, and apply it to the head. The head should be dressed in the above manner, once in every 24 hours, and not more. After it has been dressed in the above manner six or eight days, instead of the smart weed decoction, you may wash it with water in which sweet gum leaves have been bruised and soaked. I have never known the above course to fail in effecting a cure if persisted in. The patient should keep the bowels open, by the use of mild and cooling purges, and drink freely of sarsaparilla in decoction or bitters. I have known it cured by the smart weed alone, applied as above directed, and keeping it well cleansed with soap suds; I have also known it cured by cleansing it daily with soap suds, and applying a salve made by stewing the buds of the balm of gilead in sheep's tallow or hog's lard, and adding a little sweet gum rosin. Cleanliness must be particularly attended to in this disease, for without it a cure will be sought for in vain

TETTER-WORM.
Oo-coh-yok-ter-oo-ne-squaw-her

SYMPTOMS.--This disease assumes a variety of forms in different persons. It sometimes come in broad itchy spots, which run into each other, and form extensive excoriations

of the skin, or terminate in bad ulcers. Sometimes the skin thickens, wrinkles and cracks, being very tender and painful. In some persons, this complaint is most severe in winter, while others suffer most from it in summer.--This disease is sometimes constitutional and hereditary; when this is the case, it is very difficult to cure.

TREATMENT.--When this disease is in the skin only, it may be cured by very simple means. The root of the common dock, either the wide or narrow, stewed in hog's lard, and used as an ointment, will in most cases effect a cure. Puccoon root bruised, and steeped in vinegar, and used as a wash, seldom fails to effect a speedy cure.--When the case is obstinate, apply ceder oil; if it is too severe, it may be rendered more mild by stewing it with an equal quantity of hog's lard or fresh butter. I have never known this remedy to fail. When the disease is constitutional, in addition to the above external applications, give cooling purges and bitters to cleanse the blood, such as sarsaparilla, poplar root bark, sassafras or burdock. They may be taken in decoction, if preferred by the patient.

RING-WORM.

SYMPTOMS.--This disease of the skin is characterized by small red pimples which break out in a circular form, containing a thin acrid humor. It is attended with itching and smarting, which is greatly increased by over-heating the body. In a general way, the eruption does not spread to any considerable extent, but instances have occurred in which it spread over the whole body, and the skin assumed a leprous appearance. In cases of this kind, the itching is too intolerable to be borne with any degree of patience.

TREATMENT.--Ring-worms very often disappear of themselves in the course of a few days; but are apt to return in a short time. The expressed juice of either kind of dock will kill them in a short time; the juice of the walnut hull will kill them, or the juice of the inside of white walnut bark. An ointment made of equal quantities of cedar oil and hog's lard or fresh butter will also cure it. But I have seldom failed curing it, with a much more simple application. When you first discover a ring-worm, rub it well with spittle every morning before eating or drinking,

and it will soon disappear; it should be rubbed until it smarts a little.

ITCH.--[Oo-ni-tsi-lah.]

This dirty disease is infectious or catching, and is not unfrequently produced by neglecting to pay due regard to cleanliness. Some authors suppose it to be produced by a little insect which makes its way under the skin, and thus produces the eruptions and itching. I believe it to be confined to the skin, whether produced by an insect or not. Cleanliness and early attention to this disease will prevent its being communicated to a whole family. Some persons have no more regard for themselves nor others than to take this filthy disease to school and in this way communicate it to the whole neighborhood. This, however, may be prevented by the occasional use of sulphur. This disorder may be communicated by sleeping with a person who has it, or by sleeping in the same bed and bed clothes, where a person has lately slept that was infected with it. In this way travelers often catch it.

SYMPTOMS.--It shows itself first between the fingers and around the wrists in small watery pimples and gradually extends over the whole body.

TREATMENT.--Take the root of common wide or narrow dock, bruise it well and stew it in hog's lard, strain it and add sweet-gum rosin, and a small quantity of cedar oil, use this as an ointment. If it should not have the desired effect in a few days, add the flour of sulphur or pulverized brimstone. A small quantity of sulphur may be taken internally, which will prevent in a great degree, any injury by cold from its external use.

To wash frequently with a strong decoction of cedar tops or berries will generally cure the Itch, as will also a strong decoction of Virginia or black snake root, or Jamestown weed leaves or sweet-gum leaves or bark, or a decoction of buck thorn. Any of the above applications will cure the itch in a short time, provided there is due regard paid to cleanliness, without which a cure never can be affected for any length of time. The application should be made until the eruptions disappear, then the skin should be well washed with warm soap-suds and clean clothes put on. The bed clothes should also be changed, and those that were worn

while having the itch should be well washed before they are again used.

SHINGLES.

This disease is characterized by an eruption or clusters of small blisters on some part of the trunk, and spreading round the body like a girdle. They sometimes extend over the shoulder and round under the opposite arm in the form of a sword-belt. An attack of this complaint is sometimes preceded by headache and nausea, but this is not very common. The usual symptoms are, heat, itching and tingling, in some parts of the body, which is covered with small red patches of an irregular shape, and upon each of these patches may be seen very small pimples clustered together. In a short time these pimples become enlarged and are filled with a clear fluid, and the whole pimple has a transparent appearance. If the disease is not checked other clusters will appear and in a short time-extend quite round the body.

TTEATMENT.--In most cases this disease will require nothing more than the free use of some diaphoretic teas. If costiveness prevails, remove it by the use of mild cathartics, such as cream of tartar, salts or castor oil. The parts affected should be washed with a decoction of sourdock, or some cooling tea. If the contents of the bowels should be hard to remove, laxative injections should be used to aid the cathartic. All exposure to cold and damp-should be avoided, such as damp feet, damp clothes, &c. The diet should be light, and moderate exercise regularly taken.

ST. ANTHONY'S FIRE OR ERYSIPELAS.

All persons are liable to attacks of this disease--but females are most liable. The infant of a few days old, and the very aged, are equally liable to its attacks. It is generally regarded as an inflammation seated in the skin, and mostly appearing on the face, hands legs and feet, though all parts of the body are liable to its attacks. In warm climates it bears a much more inflammatory character than in colder ones. It may be produced by obstructed evacuations, such as a sudden check of perspiration, stop-page of the menses in females, or the drying up of ulcers, &c. Persons of a sanguine, irritable temperament, are most liable to its attacks; and those who have it once are much more liable to be attacked again, for that peculiar condition of the system which gives birth to it once, is more apt to occur again than if it had never existed.

SYMPTOMS.--For a few days, before it makes its appearance on the skin, great drowsiness and weakness is experienced, bad appetite, and sometimes hard breathing; more violent attacks come on with chillness, head ache, nausea and sometimes vomiting, heat and great thirst. When it makes its appearance externally, the skin becomes thick and of a crimson color and is attended with great restlessness and a burning itching sensation. The pulse is small and frequent, in a short time the skin will be covered with small red pimples containing a clear fluid. If the disease be neglected these pimples, or blisters sometimes terminate in bad ulcers, which rapidly tend to mortification.--This complaint is usually most dangerous when it attacks the face. When it attacks the face and the inflammation runs high, inflammation of the brain is to be apprehended, and should be gaurded against by all possible means.

TREATMENT.--On the first appearance of this disease, if the attack be mild, drink freely of some sweating tea, and cleanse the bowels with antibilious pills or some other cathartic. But should it put on an aggravated form and the inflammatory symptoms run high, it will be necessary to loose a little blood. Evacuate the contents of the bowels speedily as possible, and then give some diapharetic teas freely, such as sage, hysup, balm, pennyroyal, mint, or some kind of snake-root; the patient should also take the flour of sulphur occasionally, if this is not at land, pulverize brimstone as fine as possible and put it in the tea. For an external application to the skin, use fine starch or flour. If the eruption should become very painful, make a tea of red pucoon-root and use it as a wash. I have also applied a plaster made of tar mixed with a sufficient quantity of hogs lard to prevent it from sticking: this last is an excellent application, when the eruptions are so painful as to require it.

SMALL-POX
(Oo-nuh-leh-qualee)

This disease, in large cities and densely populated countries tries has proved a most fatal scourge, and has formed one of the great outlets of human life. It is contagious and extends from one country to another, spreading terror and dealing death wherever it goes. This appaling and fatal malady was unknown to the man of the forest, until their country became inhabited by the whites, and their rivers navigated by steamboats, and cities and towns were erected on their banks. It was then communicated by the whites to the Indians; it was a new disease and one for which they knew no remedy; they died indiscriminately as fast as they were seized with this king of terrors--their medicines, which they used with so much success in other diseases were tried in vain; hundreds of their tribes were hurried into eternity, while the remainder had little else to expect than soon to follow, and that in the most agonizing manner imaginable. Notwithstanding the little success they at first had in treating this alarming disease, the courage of their physicians was undaunted--they fully believed that their own happy land contained a remedy. Their former articles of medicine having failed they resorted to experiment. Every new patient afforded a new opportunity, and it was but death if the experiment should fail--and as it was death under any former treatment known by them, the poor sufferer readily consented. Those experiments were crowned with success far beyond their most sanguine expectations. They found the remedy in their own native land and in considerable abundance. This disease is now considered by them as a curable one, and that with simple and not disagreeable remedies. The art of vaccination was taught them by the whites; they use this as a preventative with great success, as may be seen under that head.

SYMPTOMS.--Small-Pox is contagious, though like measles, it seldom; if ever, attacks the same person but once. Some individuals appear to be unsusceptible to the infection of this disease, although exposed to it often through life. A few days before the appearance of the eruptions, an uneasy restless feeling and a great dislike to motion of any kind is experienced. Also chillness, followed by hot flashes and attended with slight fever. A dull heavy pain is felt in the back and in the head, accompanied with great thirst, stupor, a quick pulse, and in children violent convulsions frequently occur. About the third or fourth day, the eruptions make their appearance on the face, neck, breast and arms, in small red spots, which feel hard when pressed with the finger. These spots or pimples continue to spread until the whole body is covered with them.----They also continue to increase in size and about the fifth or sixth day, they begin to turn white on the top and feel very painful. About this time the face swells and the features appear changed, the eyes are frequently closed by the swelling and the quantity of spittle greatly increased The throat swells and is attended with hoarseness and difficulty of swallowing; by the tenth or eleventh day, the pimples have increased to about the size of a common pea, on the top of each pimple or pustule there will be a small black spot, whilst the remainder of the pimple or pustule will be filled with yellow matter. In a short time the swelling in the face in some degree subsides, and a swelling in the hands and feet follow. The pustules gradually become rough, break and discharge an offensive matter. When these pustules are large or long in drying, they leave deep holes or scars in the skin. Some writers divide this disease into two species, distinguished as the *distinct* and *confluent.* The above are the symptoms attending the distinct. The *confluent* is more rapid in its progress than the *distinct.* It assumes a *typhoid* or nervous form, but as our mode of treatment is similar in both forms of this disease, the distinction need not be minutely drawn.

TREATMENT--First give a large dose of antibilious pills and aid their operation by laxative injections. (See under the head of clystering.) After the contents of the bowels have been thoroughly evacuated, let the patient drink freely of a tea or decoction of water big-leaf, called by the Indian, *Oo-cah-lo-ka-guah-ah-my-yeah.* Also wash the patient frequently all over in a tea of the same; the bowels must be kept open by the use of antibilious pills, and the free use of the tea continued until relief is obtained. For direction to prepare the antibilious pills refer to that head. Also, for a full description of the Water-Big-Leaf to that head.--This valuable medicine is considered by us an infallible remedy for this alarming scourge, the Small-Pox, depriving it at once of its former terrors and fatality.

VACCINATION.

Vaccination is the introduction or insertion into the arm by means of the lancet, of the matter by which the *Cowpox* is produced in the human system. About fifty-five or sixty years ago, vaccination was discovered to be a preventative of Small-Pox. This valuable discovery was made in England, and has since been made known to the different portions of the globe. It was communicated to the Indians by the English, and is now used by them as a preventative of this dreadful scourge, and however easy may be the cure, yet the preventative is to be preferred.--The matter should be taken before the ninth day, or it will be too inactive to produce the desired effect.

When vaccination takes proper effect on the system, we believe it to be a safe preventative, and one that may be relied on without the least anticipations of danger. The proper mode of communicating the vacine matter to an individual, is by scratching the skin with a needle or lancet, until a small quantity of blood appears; then put a small quantity of the matter in the scratch or incision. If the matter is good, it will appear transparent; on the contrary, if it has become opaque, it should not be used. This matter is sometimes brought to this and other countries on thread: when this is the case, make the incision as above directed, and put a piece of the thread which contains the matter in the incision, and confine it there. The Cowpox sometimes fails to take the proper effect on the system; in this case it only produces a slight redness in the arm, and is attended with no other disagreeable feeling. On the contrary, when it has the desired effect, it generally produces a slight head ache, pain under the arm, and some fever,--these symptoms generally pass off in the course of a few hours, without medical aid. The proper place for inserting the vacine matter, is on the arm above the elbow, and when it takes the desired effect on the system, it is easily known by the appearance of the lump or pustule, which rises at the place where the matter was inserted. If a pustule or pimple should arise of a full and oval form, with a dent or indentation in the centre, and contain matter, the vaccination has had the proper effect. These pustules generally make their appearance about the third day--about the eighth day the pustule is completely fromed.

In a short time the disagreeable feelings subside, and the pustule generally disappears.

CHICKEN POX.

This is an infectious disease, and never attacks the same person but once through life. It is characterized by pimples or pustules, on the skin which bear some resemblance to those of the small-pox--though of a much milder form. The appearance of the pimples or pustules is usually preceded by slight fever, attended with chillness and stupor. or drowsiness--a pain is felt in the head and back, great thirst, restlessness and a quick pulse. About the third or fourth day, the pustules become filled with a watery substance, which, however, never becomes yellow as is the case in small pox. A few days after the pustules become full, they dry away--leaving a crust or scab over each pustule--which sometimes leaves a scab when it comes off, though not very often. A malignant form of chicken pox is called swine pox. The same treatment will answer both, only it should be followed with more promptness in the latter, than in the former case.

TREATMENT.--The bowels should be kept open by the use of mild cathartics, such as castor oil, cream of Tartar. American senna, or any other cooling laxative that may be preferred. Let the patient drink freely of some diaphoretic teas such as Mountain dittany, black dittany, sage balm, hysop, pennyroyal or either kind of snake root; and avoid all exposure to cold or sudden changes, in order to prevent the eruption from striking in and producing sickness. Where the disease assumes a very aggravated form, the water big leaf may be used as directed for small pox. By the observance of the above rules, no danger should be apprehended from the swine or chicken pox.--The diet should be light and nourishing.

MEASLES.

This is a contagious or catching disease, but like many others it attacks the same person but one time during life.

SYMPTOMS.--Between the third and ninth day--some authors prolong the time to the fifteenth day--after the infection has been received, it produces sickness at the stomach, stupor, dullness, great thirst, frequent sneezing, as if taking a severe cold, a short dry cough, redness and watering of the eyes, and the running of a watery mucous from the nose. In a short time after these symptoms, the eruption generally makes its appearance on the face and neck in small red pimples, which soon spread over the whole body. The fever and cough does not abate in the measles on the appearance of the eruption as it does in small pox, but continues until the eruptions begin to disappear, which is generally in about three or four days after it makes its appearance. In some instances the cough and fever has been known to continue for several days after the eruptions had entirely disappeared, this however is not very common.

TREATMENT.--The treatment for measles in ordinary cases, when the patient in other respects is in health, is simple--nothing more is necessary than to drink freely of some diaphoretic or sweating tea, and keep the bowels open by injections or mild cathartics. But in no disease with which I am acquainted is there more serious consequences to be apprehended from taking cold, than in Measles.

The above treatment will in most cases cause the Measles to strike out in due time. But should the above mild means fail to bring the eruptions to the surface, and the fever is high, bleeding will be necessary, and a purge of castor oil or cream of tartar will also be necessary. If there is an inclination to puke, an emetic will not be improper. After bleeding and puking, if the occasion requires let the patient sit over a steam of ceder tops, and drink freely of a tea of the same; bathe the feet frequently in warm water; continue this course until the eruptions make their appearance on the skin. When the cough is severe, and the throat very sore, suck the steam of hot water or vinegar from the spout of a coffee pot. Cold drinks must not be indulged in, warm sage, balm, hysop, or pennyroyal tea will be proper for their drinks. Spirituous liquors must not be used in this disease under any circumstances, unless it is at the commencement of mild attacks. When the throat swells and inflames, treat it as directed for malignant sore throat. After the abatement of the inflammatory symptoms, if the patient is very weak, give tonics to increase the strength of the system.

MUMPS--(Te-le-g-nah-tsi-luh-no-tis-say.)

SYMPTOMS.--It usually commences with a slight fever, head ache, a stiffness of the neck, and a swelling under the lower jaw, on one or both sides. On attempting to swallow a severe pain is felt precisely at the point of the lower jaw, and extending to the ear; the swelling increases, and by the fourth or fifth day, the part is greatly swelled.

In some instances, the color is but little changed; in others, the skin assumes a red appearance--about the fifth or sixth day, the swelling begins to subside, and by the seventh or eighth day, the swelling is nearly gone. If cold has been taken, during the above symptoms, the swelling as it leaves the jaws, is apt to settle in the testicles, if the patient be a male, and in the breasts, if a female. It is infectious, but I believe, it never attacks the same individual but once. Some writers assert that it has been known to occur in the same individual more than once; others say that it will, in some instances, affect one side, and at some future period, when the disease prevails, the other side may be affected. But as neither of these latter cases has ever come within my own knowledge, I am unable to say any thing in reference to the correctness or incorrectness of these statements.

TREATMENT.--The bowels should be opened by the use of castor oil, cream of tartar, American senna or some other mild cathartic, aided by injections, if necessary, and avoid exposure to cold, such as damp or cold air, wet or damp feet, damp clothes, &c. Flannel should be kept around the jaws, to prevent exposure to the atmosphere--if this is not at hand, a thin bit of wool will answer equally well. It should be confined around the jaws by means of a handkerchief. If the jaws become very painful, apply a plaster made as follows. Put beeswax into an oven of warm water, and when the wax is melted, take a cloth large enough to cover the swelling in the lower jaws as the case may be: dip the cloth into the oven, and when taken out the wax will adhere to the cloth; apply this to the swelling as warm as can be borne. If this does not abate the swelling in two days, apply a plaster made in the above manner of bees-wax, tallow and red pepper. A free use of diaphoretic or sweating teas will be found highly beneficial. Such as pennyroyal, hysop, dittany, &c. &c.

BURNS AND SCALDS.--[Oo-ne-log-yer-suk.]

These painful accidents are often the offspring of negligence, and when first received, are very painful. It is very desirable, therefore, to have a remedy, at hand, that will at once relieve the pain and extract the fire. Nature, in her liberal dispensations of blessings on mankind, has not failed to provide a soothing and effectual remedy for this painful emergency. On receiving a scald or burn, immediately plunge the part into cold water, and keep it there until the fire is extracted, which may be easily known by taking it out of the water, if the fire be not out it will smart and pain you as a burn does, as soon as it is out of the water and exposed to the atmosphere; but if the fire is out, no such pain will be felt. If the burn or scald is so situated that you cannot immerse it in water, immediately exclude the atmosphere by means of cloths, which should be several piles thick over the burn, and keep them wet by pouring cold water on them; the air should be excluded twelve hours by means of the cloths, which should be kept wet with cold water. If this is properly attended to, as soon as the scald or burn is received, it will in nine cases out of ten prevent it from blistering, and consequently prevent the formation of a sore or ulcer from the burn. To prevent the patient from taking cold from the application of cold water; let him drink pepper or some other sweating tea freely, or warm toddy, will answer equally well. To small burns, where the skin is broken, apply a salve made by stewing equal quantities of pulverized bark of elder, and the white part of hen dung in hog's lard.--Where a person is badly burnt on the body, kill a cow brute, and cut it open as soon as possible; the maw or paunch should be opened also; the patient should be put in it as soon as possible, and remain there until the animal becomes cool, then take him out and wash him off with cold water, and dress the wound with healing salve, and wrap him in comfortable cloths. The patient should drink freely of some sweating tea and warm toddy, to keep up the internal heat, and if possible produce a determination to the surface, or in other words sweating. If the burn should produce a chill, drink freely of sage, balm or pepper tea while the chill continues. Then take antibillious pills to evacuate the contents of the bowels as soon as possible, at the same time treating the wound as above directed. I was, says Dr. Foreman, "called upon to attend a boy whose body was so badly burned as to render the motion of the *abdominal viscera,* (which means the intestines or guts) very visible through the thin membrane which covered them. This case was thought to be beyond the reach of remedies, by all who saw it. I had a cow brute slain immediately, and the boy put in it, while it was warm and bleeding; when it became cool, he was taken out, and treated as above directed. He lived and recovered in a short time. This is the most horrible and alarming burn I have ever had, and the only one in which I resorted to that mode of treatment. But it has been the custom of my people, (the Indians,) for many years, and has been attended with admirable success. Dr. Gunn mentions Turner's cerate, as being one of the most soothing applications that can be made to a bad burn on the body. I have never tried it myself, but feel disposed to place implicit confidence in the statement of this gentleman, he being a very successful physician of the old school. It is easily prepared, and if it is possessed of the active medical virtues ascribed to it by Dr. Gunn, it should be kept in every family for immediate application. The Index will refer you to its proper head, where may be seen the manner of preparing and using this valuable salve.

FRACTURES AND DISLOCATIONS.
[Oo-nah-tuh-log-sah.]

When a joint is dislocated or a bone fractured, apply cloths wrung out of a tea of ivy as hot as the patient can bear, the vessel containing tea should be set near and the tea kept constantly pouring on the cloths for fifteen or twenty minutes--if the ivy is not at hand, hot water will answer--then take the cloths off, and if it is a dislocation, pull the limb steadily until it returns to its proper place, after which pour cold water on the joint for a minute or two in order to prevent a second dislocation. If the joint should be painful apply brown paper wet with vinegar or camphor, or a poultice of sprain-weed. In case of fractures or broken bones, when the cloths are taken off, the bone or bones should be placed in their proper situation by some skillful person, and confined to their place by splints. If it is the leg or thigh bone, it should be laid on a box, and secured by bats of cotton, and the foot kept nearly level with the knee. The bandage which confines the splints should not be drawn too tight, as is the custom with many persons who profess to understand replacing and bandaging broken bones, this obstructs the circulation and greatly retards nature in effecting a cure. After it is splintered apply a poultice of sprain-weed or young elder root pounded fine and mixed with cold water and wheat bran to the consistency of a poultice; the part should also be wet frequently with cold vinegar or camphor. The splints may be taken off after the fifth day, once every day by some careful person and the limb held with great care in the hands in order to give the patient some rest, and again put on. This must, however, be done with the greatest caution imaginable, as the bone will not re-unite until between the ninth and twelfth day, the splints should not remain off before the fifteenth day.

WOUNDS AND CUTS.--(Oo-nah-tah-leh-ger.)

Most ordinary cuts require but little attention, except binding up with a cloth or bandage, and occasionally wetting the cloth with cold water, until it begins to matter, then apply healing salve. If inflammation should take place, reduce it with a poultice of beach, dog-wood, or either kind of oak-bark; or tar plaster will in most instances effect a speedy cure, without other remedies. But when the cut is large and bleeds freely, wash off the blood with cold water, cleansing the wound of all dirt or filth, then draw the edges of the cut together, and bind it up carefully, and occasionally pour cold water on it; this should be done as often as it feels hot. The patient should at the same time drink some sweating tea: if the blood should flow rapidly, use styptics. If the incision or cut be very large, it will be necessary to confine the edges together with a few stitches, or by the application of an adhesive plaster. When large arteries are wounded, and the bleeding cannot be stopped by styptics, a ligature is necessary for this purpose; prepare a cushion or roll up a handkerchief in the form of a cushion, and place it on the artery above the wound; then draw a ligature around the cushion and limb, tight enough to stop the blood, then tie the artery. It may be readily known when an artery is wounded by the blood, which does not flow in a continual stream, but by spirts. Bleeding is often stopped by raising

the wound above the heart or head, binding it up tight, and pouring cold water on it freely. If the wound inflames, apply poultices to reduce the inflammation.

HOOPING-COUGH.

All persons who have never had this complaint are liable to it. It attacks but once through life, and is contagious, or catching, and epidemic.

SYMPTOMS.--It begins nearly like a slight cold, but is attended with more weakness, head-ache, hard breathing, sneezing, hoarseness, with a little cough, which gradually increases until the face becomes bloated and turns purple, the eyes swell and become prominent.

TREATMENT.--Keep the bowels regulated by mild and cooling purges, and give a tea of *Culsay-tse-e-you-see.*--(See proper head,)--where it is fully described. A little laudanum or paregoric given at bed time, will not be a miss in severe cases.

PART III

CHAPTER 1.

DISEASES PECULIAR TO THE UNIMPREGNATED STATE.

Previous to the age of puberty, the female is scarcely subject to any disease not common to both sexes, but when that period arrives, they are not only liable to all the ordinary diseases to which men are exposed, but in consequence of their sexual organization, they are also subject to many diseases peculiar to themselves. The organic machine in women is more complex than in men. and the functions performed by these organs are easily deranged, from which diseases of an inveterate and dangerous character often arise. Woman is liable to painful irregularities in her menstrual discharges, which may not produce ill health at the time, but lays the foundation of lasting and dangerous diseases. These irregularities must be early and properly treated, or they will involve the general health, rain the constitution, and bring on dropsy, consumption, or some other fatal disease. After puberty, almost every stage of female existence is subject to some complaints peculiar to itself, as well as to those diseases common to all. These will be treated of under their proper heads.

SECTION I.
MENSTRUATION.

Menstruation is that periodical discharge which takes place from the womb, commonly called menses or courses. The term Menses is derived from the Latin word *mensis,* which signifies a month, because in healthy women, who are neither pregnant nor giving suck, this discharge generally flows regularly at intervals of a lunar month, or about twenty eight days. With some, the intervals are a day or two longer, whilst with others, it is a day or two shorter.

The period at which Menstruation commences, depends very much upon the climate, constitution, and mode of life. In warm climates, the menses often appear at eight or nine years of age. In temperate climates, they generally make their appearance at from twelve to fourteen, and in cold climates, they do not appear until the eighteenth or twentieth year, and even twenty-fifth year.--
There will also be a difference even in the same-climate as to the time of Menstruation, which depend upon the constitution and passions. Those who have a rapid growth of body and development of the organs, with warm passions, will have an earlier discharge of the Menses than those who are different in these respects.

The time required for the Menstrual purgation, at each periodical return, is from three to six days. The Menstrual fluid appears to be a regular secretion from the womb, which in its appearance very much resembles blood, and its regular discharge, at the proper intervals, are important to the health of a woman, from the time of its appearance until the age at which it should entirely cease, except during pregnancy, and during the period of giving suck.

The period during which the menses continue, until they cease entirely, varies according to the time of their commencement,--the time being generally about double that which elapse previous to their commencement. Whenever the menstrual discharge makes its first appearance, it announces puberty, and not maturity of the generative organs, which renders them capable of performing the functions for which they were created; and when this discharge ceases, or leaves off entirely, it announces the inability of the generative organs to perform their peculiar functions. Both these periods are critical with women, and much depends upon the precautions in avoiding exposure to cold and wet, or overstraining in lifting, working, &c. Many girls have their discharges without inconvenience, while others suffer considerably when the period is about to come on; such as great restlessness, slight fever, head-ache, heavy dull pain in the small of the back and lower part of the abdomen, swelled and hardened breasts, &c. The appetite becomes delicate, the limbs tremble and feel weak, the face becomes pale, and there is a peculiar dark streak or shade under the eyes. When these symptoms and feelings occur, every possible care should be taken to avoid cold, damp, &c., and everything should be done which would assist nature in bringing forward this discharge. This is a critical period of life, and much indeed depends upon the result. The greatest possible precautions should be used to prevent the girl from taking cold at this time, because by very slight exposure, nature may be prevented from performing this very important office; by the failure of which, some of the most fatal female diseases are produced. Exercise should be taken on horseback, or indeed any exercise that will give free circulation to the blood. The emotions and passions of the mind, ought to be particularly attended to; a cheerful disposition should be produced and kept up; and every effort should be made to banish grief, despondency, or any of the depressing passions, which if indulged in, will not fail to have a powerful effect in preventing the due discharge of the menses, or courses. About this time of life, girls should not be allowed to get wet, wear damp clothes, sleep on damp beds, walk in grass wet with dew or rain, nor walk bare foot on cold or wet ground. You should also avoid everything that will have a tendency to injure the digestive powers, and particularly costiveness, or being bound in the bowels, loss of sleep, exposure of any kind, tight lacing, &c. When the first symptoms of menses make their appearance on young girls, they should use all mild and gentle methods of courting nature to the performance of her office, by sitting over the steam of warm herbs, bathing their feet and legs in warm water as high as the knees, and drinking of warm pennyroal tea. These means should be used immediately before going to bed, so that a gentle moisture or sweat may be produced on the skin, which generally causes the menses to flow. A little care and attention on the part of the parent at this period, may be of lasting benefit.

The menstrual discharges on their first appearance, are generally in very small quantities and somewhat irregular as to time, but by attending to the simple course which I have laid down, they will gradually increase and flow monthly.

SECTION II.
RETENTION OF THE MENSES.
(Tsa-his-lee-ah-nah-tah-gah-ta-gee.)

By retention of the Menses, is meant the retaining or keeping of the menstrual fluid, after the period of life has arrived when this discharge should take place.

When girls arrive at the age of puberty, the menstrual purgation is essential to their health, and if it does not take place, there will be headache, loss of appetite, weakness of the limbs, a peculiar paleness of the face, accompanied with a sinking of the spirits, hysterical affections and other derangements of the general health. When girls have arrived at the age when this discharge should appear, nature generally gives an indication of the by-pains in the back, hips and loins, a sensation of weight, fullness and heat in the pelvis, attended with a forcing or heaving down. If no discharge takes place, these symptoms sometimes occur periodically, until continued bad health is produced, and will ultimately seat some fatal disease, if not counteracted by the aid of remedies and prudent management.

TREATMENT.--The vegetable kingdom affords many valuable articles for this painful and extremely dangerous complaint, as is fully shown in Materia Medica. By referring to that part of this work, the reader will find a full description of many valuable roots and herbs for this disease.

The patient should take exercise in the open air in fair weather, but she must carefully avoid damp air, night air, walking in dew, or going barefoot in cold or wet places; exercise on horseback would be best. She should keep the bowels regulated by the use of mild and cooling purges, bathe her feet frequently with warm water, and drink freely of some diaphoretic or sweating tea, just before going to bed. She should also drink daily of bitters, composed of ginger root, star root, rattle root, Sampson snake root, wild cucumber bark, or common tansy. The above roots and barks may be used alone, or several of them together in spirits, as the patient may prefer.

SECTION III.
IMPERFORATION OF THE HYMEN.

The Hymen is a thin membrane, found at the mouth of the vagina, and, in general, it partly closes the entrance of the vagina. Some instances have occurred, in which it entirely closed the vagina, and was so strong as not to give way at the proper time, for the commencement of the monthly courses or menses; in which case, it must produce, at no mature age, serious, and unless removed, fatal consequences. An imperforation of the Hymen is attended with no inconvenience, until the monthly purgations should take place. If this membrane be imperforated, and the menstrual fluid being regularly secreted, must accumulate, both in the vagina and the womb, as it can find no outlet through the Hymen. In some intances the quantity accumulated has been so great as to subject the unfortunate sufferer to the suspicion of being pregnant. At each return of the menstrual period, considerable pain is experienced by the patient, and as these pains greatly resemble those of labor, in cases where the enlargement of the abdomen was considerable, they have been mistaken for labor. After these pains continue for some time, they cease, and do not recur until the return of another menstrual period.

When the menstrual fluid has been contained, in consequence of the imperforate state of the hymen, it assumes a dark tarry appearance, and unless its evacuation be procured by opening a passage through the hymen, serious injury to the health will be sustained, or some fatal disease produced.

The only means of removing the difficulty, is by making an artificial perforation or opening, through the membrane. This operation is quite simple, and may be performed by any sensible female friend, with a lancet.--Care should be taken not to cut any of the contiguous parts, and no particular danger is to be apprehended from the operation, nor is it attended with much pain, as the membrane does not possess great sensibility.

SECTION IV.
SUPPRESSED OR OBSTRUCTED MENSES.
(Tsa-yoh-tus-let-ah-nah-tah-gah-ta-gee.)

When the menses have made their appearance, they are liable to be obstructed by cold, &c., this is called suppressed or obstructed menses, and is attended with greater or less misery according to the state of the system at the time this obstruction takes place, and more particularly if any other part of the body is laboring under disease.--The bad effects of taking cold do not always show themselves immediately, but they generally become manifest, after the repeated return of the period at which the menstrual discharge should take place, if the obstructions be not removed. Women that are in good health, may not experience any inconvenience for some months, that is, until the periodical returns shall have passed several time without the necessary discharge.

But such is the sympathy existing between the womb and other parts of the system, that the general health will be effected, and sometimes the most incurable diseases are firmly seated in the system, by neglecting to remove the obstructions at an early period. Hysterics, depression of spirits, sickness of the stomach, pains in the head, back and bowels, coldness of the hands and feet, flashes of heat over the body, spitting blood, bleeding at the nose, colics, a dry short cough, pains in the abdomen, a hard, quick pulse, a hot skin, and a burning sensation of the palms of the hands, and bottoms of the feet, are symptoms generally met with, when the menstrual discharge has been obstructed long enough to produce some disordered state of the womb. When the last above named symptoms occur, they indicate great danger from the consumption, and unless relief is immediately had, that fatal disease will be confirmed--negligence at this critical period, will, in most cases, be followed by fatal consequences.

TREATMENT.--As soon as it is discovered that the monthly purgation is obstructed, and it is believed that cold and not pregnancy is the cause, you should take measures to remove the obstruction, which is much more easily done in an early stage than at a more advanced period. About the time the menses should flow, the patient should drink freely of a tea of tansey, dittany, balm, rattle-root, penny-royal or some sweating tea. The feet should be well bathed before going to bead, and every mild and gentle means should be used to produce perspiration or sweat. If these remedies should fail, the patient should drink bitters, as directed for Retention of the menses. And at each periodical return when the menses should flow she should drink sweating teas and bathe her feet as above directed. She should also sit over a steam of young pine tops, cedar tops or spruce pine tops; while over the steam she should take a strong decoction of seneka snake-root and pleurisy root in table spoonful doses every ten minutes. When the patient leaves the steam, she should cover up warm in bed, and continue drinking some sweating tea for a length of time, in order to keep up a free perspiration. Great care must be taken to cool off by degrees after the above course, as there is danger to be apprehended from taking cold in case of neglect. The treatment for suppressed or obstructed menses, and retention of the menses, is nearly or quite the same--what is good in one case is also good in the other, and either of these complaints may be overcome by mild and gentle means, in ninety-nine cases out of a hundred, if taken in due time and perseveringly attended; on the other hand, neglect is invariably followed by serious, and sometimes fatal consequences. Where there is irritation of the nerves, some of the articles in the class of anti-spasmodics may be combined with the bitters and teas, such as the moccasin-flower root, ginseng, asafoetida. &c.

SECTION V.
PAINFUL MENSTRUATION.

THIS painful malady is often met with in our climate, and is often not only accompanied with great sufferings but is frequently obstinate to cure. The causes of this complaint are supposed to be taking cold during the flow of the menses, or shortly after abortion.

The quantity of menstrual fluid discharged is generally small, and is accompanied with severe, bearing down pains, similar to those of labor, the pains come on at intevals, and continue until small clots of blood are discharged, after this discharge some ease is experienced until a fresh production of this substance is to be expelled when there is a return of the pains. Women afflicted with this complaint seldom bear children until cured.

TREATMENT.--The patient should first take a dose of anti-bilious pills or some other cathartic to cleanse the bowels. About the time the menstrual discharge is expected, she should drink freely of tansey or some worm wood tea, and set over the steam of young cedar or pinetops; sweating teas should be drank freely just before going to bed; the patient should also make a daily use of some laxative tonic in bitters, or some of the preparations recommended under the head of bitter laxative tonics, in the dispensatory. In many instances, bitters of Columbo root and Burdock root will answer admirably well. At the time of menstruation, when the pain is very severe, the patient may take a teaspoonful of paragoric or Bateman's drops in her tea. A tea of chamomile, either the herb or flower is very good; as is also a tea of the common garden marigold flower, or a tea of winter clover--one berry.

SECTION VI.
GREEN SICKNESS.

WHEN the menses or courses have been retained or stopped for any length of time, and the whole system becomes diseased from want of this discharge so necessary to the health of every female, it terminates or ends frequently in what is called chlorois or green sickness. In this disease the skin turns of a pale yellow or greenish hue, the lips become pale or of a purple color, the eyes have a dark or purple tinge around them, there is frequent sickness without knowing the cause, on making the least exertion the heart palpitates or beats, and the Knees tremble--the cheeks are frequently flushed as in consumption, the mind is feeble and the woman seems to lack the power to attend to her domestic affairs, the feet swell and the whole system seems to sink under great debility or weakness.

TREATMENT.--In this disease the patient labors under extreme debility; therefore, tonics and strengthening medicines are required. If the bowels are costive, give some laxative until their condition is changed, as soon as the contents of the bowels have been evacuated by the use of laxatives, commence giving the chalybeate pill night and morning, say two small pills for a dose--if there pills cannot be conveniently had, give iron or steal dust in same way. The patient should also use the hepatic pill once a day, two for a dose, and drink bitters of star-root, columbo-root, wild cherry-tree bark, rattle-root, or any of the bitters recommended for retention of the menses. The patient should bathe the feet, and drink some sweating tea, every night before going to bed. All exposure to cold, damp and wet, must be avoided. The patient should take moderate exercise, but avoid fatigue. The diet should be such as the stomach will easily digest, but let it be nourishing.

SECTION VII.
PROFUSE MENSTRUATION.
[Oh-ne-ta-sha-ne-tsa-ne-yoh-oo-lah.]

The menstrual discharge may be too profuse, either from its too frequent recurrence, or from the great quantity discharged, when recurring at the proper periods.

The causes are too great a determination of blood to the womb, or in other words too great an action in its vessels. This over quantity, or large discharge, generally takes place in delicate women, particularly those who take but little exercise, or those who sit a great deal.

TREATMENT.--The patient must be kept cool and quiet and spend as much of her time in bed as possible, with her head very low. A decoction of cumfrey root may be used to great advantage, it must be drank cool. A tea of princes feather queen of the meadow, and red root are all good. But if these remedies should fail, give a tea of a decoction of Oo-na-tah-cah-see-le-shee. I have never known this article to fail. Any astringent tonic is good in profuse menstruation. After temporary relief is obtained, the Chalybeate pill or some strengthening medicine must be used to improve the general health.

SECTION VIII.
CESSATION OF THE MENSES.

Cessation of the menses or courses means an entire stoppage of this discharge, or a change of nature, when the female has arrived at that period, in life, when these organs become incapable of performing their peculiar functions. This change usually takes place between the forty second and forty seventh year, though in those of delicate constitution, it stops before that period, and in those of robust condition, it sometimes continues later--it is a critical and extremely dangerous period of a woman's life and notwithstanding thousands pass through it without experiencing any inconvenience; it is a period which requires particular care and attention. All exposure to cold and damp must be scrupulously avoided, and particularly wet feet, and remaining long on the damp ground. Sudden changes of dress and every thing that produces sudden revolutions in the bodily system from extremes of heat and cold, and dampness. But not attending to the above precautions, you will be sure to lay the foundation of diseases of a multiplied and stubborn character, which will be sure to embitter and destroy the remainder of your days, let them be many or few.

The cessation usually takes place gradually. They first diminish in quantity, and become more irregular, until they return no more. Strict attention to temperance and exercise, so as to preserve the general health, and promote the free exercise of all the other functions of the body, is necessary.

If any disease should ensue, treat it according to the directions laid down under its proper head.

SECTION IX.
WHITES AND FLOUR ALBUS.--(Oo-na-yah.)

Whites or Flour Albus is an unnatural and white colored discharge from the birth-place, and is produced from various causes, such, for instance, as the powers of the womb being impaired by severe labors, repeated miscarriages, getting out of bed too soon after child-birth, or by taking cold at this time, or any other time when the menses are about coming on; this disease is sometimes brought on by fatigue, or weakness produced by general bad health. Women of weakly and delicate constitutions, such as take but little exercise, and those who have had many children are much subject to flour albus or whites; in some instances this discharge makes its appearance monthly, instead of the natural menses or courses. This is generally the case where the woman is laboring under suppression of the menses or some derangement of the whole system. The most aggravated form of this disease, and the mildest form of clap in females bear a strong resemblance. Some writers on this subject say, that they may be distinguished by the discharge in clap producing a scalding and burning sensation, whereas no such feelings are produced by the discharge in Flour Albus or Whites--this, however, is not the case, for the discharge in Whites often produces itching, uneasiness, great heat and scalding of the parts. In clap, there is a swelling of the parts and the scalding sensations increase in severity much faster than in Flour Albus.

The whites are called by this name, because the discharge resembles the white of an egg. There are several stages of the complaint, and between the mildest and severest form, if permitted to run on, it will entirely destroy the constitution and seat some incurable disease on the system. The complexion will change first to a pale sickly color, and if the disease is permitted to run on, it will at last assume a sickly, greenish hue, and the lips become purple; at this state seek for a remedy under the general head of green sickness.

Whites and green sickness are sometimes produced by the falling of the womb. When either of these complaints is caused by the falling of the womb, look at page 200 for a remedy.

TREATMENT.--The bowels should be emptied with antibilious pills, and costivenesss prevented by the occasional use of the same in small doses, or by the use of gulver's root, castor oil or cream of tartar. After the bowels have been emptied, give the chalybeate pill, night and morning, two common sized pills for a dose, and make a constant use of the tea of Oo-wa-sco-you, called by the whites blue flag or gleet root, a pint of this tea should be drank each day tolerably strong. Particular attention should be paid to cleanliness of the parts. They should be washed frequently with warm water, and some astringent article injected up the birth-place, such as oak oose, brier root tea, wild alum root tea, &c.

CHAPTER II.
DISEASES OF THE PREGNANT STATE.

Pregnancy, though not a disease, is often attended with diseases peculiar to that state which are very troublesome. The diseases commonly attendant on Pregnancy, are not of a very dangerous character; yet some of them produce the sorest ills that afflict the female race. "Many a female appears to have the curse pronounced upon Eve fully veryfied in her own case. Sorrow marks her for her own from the time gestation commences until the period of her deliverance." But this is not always the case: some women enjoy an unusual portion of health while in a state of pregnancy, "but these favorites of heaven are like angels visits, few and far between." The system during pregnancy experiences an increased susceptibility of disease.

SECTION I.
SIGNS OF PREGNANCY.

Young healthy women, whose monthly terms appear regular, may commonly know when they are pregnant by the terms or menses not returning at the proper period; there is often sickness and vomiting, particularly of a morning: it is frequently attended with heart-burn and sourness of the stomach, loss of appetite, craving for food which before was disliked, and often a particular dislike to diets as had previously been held in high esteem. The face becomes pale, the features sharp, the waist grows more slim and lank than usual and continues so for some time. The breasts become more full and the rose-colored ring around the nipple becomes darker. Toothache is frequently an indication of pregnancy. The rising of the naval so as to become flat and smooth with the belly, may be considered almost a certain sign of pregnancy.

During pregnancy, some women become peevish dull, and gloomy, and others are more lively and agreeable than usual. The pulse during pregnancy, is considerably quicker than common, and there is frequently a dizziness or swimming in the head. Pregnancy never does exist without some or all of the above named symptoms, yet the most of them may exist without pregnancy. There is but one certain sign of pregnancy, which is the *motion of the child felt by the mother;* between the end of the third month and the beginning of the fifth the motion of the child can be distinctly felt by the mother, which is called quickning, and when quickning is felt it is a certain sign of Pregnancy.

SECTION II
SICKNESS OF THE STOMACH AND VOMITING.

Very few females escape this distressing and common attendant at the earlier stages of pregnancy. If the vomiting is not severe, it will do no injury, but if it should be very severe and produce considerable debility, means should be used to lessen its severity or stop it entirely.

TREATMENT.--The bowels should be kept regulated by the daily use of laxatives of a coolling nature, cream of tartar will answer well for this purpose. The gentian and columbo root taken together or alone are excellent, they may be taken in bitters if preferred, to which may be added the essence of peppermint--or they may be taken in tea or a decoction and the peppermint added; it should be permitted to get cold before it is drank, ginger may be put in the bitters or decoction, with advantage. If any particular kind of food be craved it should be procured, as the gratification of the capricious appetite seldom fails in diminishing the severity of the symptoms. A cup of ginger or mint tea will often give relief. The principle remedy however, and the one most to be relied on, is keeping the bowels open by cooling laxatives and clysters. More than halt the diseases which arise during pregnancy, are more or less occasioned by a costive state of the bowels, and every pregnant woman should bear in mind the vast necessity of keeping the bowels so regulated as to have a stool daily, whenever she falls short of this she endangers her health.

SECTION III.
CRAMP.

Cramp, with some, is an early attendant symptom of pregnancy, but it generally comes on about the fourth month of pregnancy, it is commonly most troublesome at night while in bed. It attacks different parts of the body, and is generally most severe during the latter stages of pregnancy.

TREATMENT.--When the Cramp comes on get out of bed immediately, stand a few minutes on the coldest rock that can be procured. This will give present relief. To prevent its return, keep the bowels in good order with purgatives or injections, and confine the flour of sulpher or powdered brimstone around the legs by means of a garter or belt. If the attacks are frequently and violent, in addition to the above, rub the parts with the essence of pepper, and if the patient be of full habit draw a little blood.

SECTION IV.
PAIN IN THE HEAD AND DROWSINESS.

Unpleasant sensations of this kind, very frequently occur during pregnancy; they are in most instances occasioned by the blood vessels being too full; but sometimes in delicate, weakly women, they arrise from an opposite cause, such as a want of a due circulation of the blood which induces debility or weakness.

TREATMENT.--If the woman thus afflicted be fleshy and strong, draw blood from the arm and keep the bowels open by the use of some laxative medicine, such as anti-bilious pills, castor oil, cream of tartar, rheubarb, &c. But if she be weakly and delicate, bleeding will be highly improper. She should take moderate exercise, but avoid fatigue by all means. The bowels should be regulated by the use of very mild laxatives or injections. She should drink freely of columbo and spikenard bitters, and bathe her temples frequently with spirits in which camphor has been dissolved. The warm bath is excellent in cases of this kind.

SECTION V.
SWELLED LEGS.

This swelling is produced by the weight of the womb pressing on the vessels which return the fluid from the lower parts of the body. The womb is greatly enlarged during pregnancy, and in the advanced stages of pregnancy these swellings frequently give great pain.

TREATMENT.--Let the woman go to bed and remain as quiet as possible, if she be stout let her loose a little blood, and regulate the bowels by the use of mild and cooling medicines, such as cream of tartar, rheubarb, &c. Women afflicted in this way should spend as much of their time in a lying posture as convenient. There need be no danger apprehended from thes swellings, although they often prove troublesome.

SECTION VI.
HEART-BURN.

Very few women escape this distressing complaint during pregnacy: it generally arises from acid in the stomach.

TREATMENT.--If heart-burn is attended with sickness at the stomach and a constant hawking up of a tough phlegm, it will be necessary to cleanse the stomach with a gentle emetic, such as ipecac or indian physic. But if it is accompanied with a hot sour taste in the mouth, and a belching up of sour water, it may be relieved by the use of weak lye or lime water, or by a tea-spoonful of magnesia in a cup of cold water, or it may be eaten if preferred.--Ground ginger is also good for Heart-Burn; it may be taken in half tea-spoonful doses as often as necessary or the roots may be chewed at pleasure. Cinnamon bark is also good for the Heart-Burn. Slippery-elm bark powdered and taken in cold water, is an excellent article for this distressing complaint. Comfrey, either the gardeu or wild, will generally give speedy relief. The slippery-elm bark and comfrey will act as an aperient, and will probably afford the most permanent relief of any of the above named articles, for costiveness should be strictly avoided by persons afflicted with Heart-Burn.

SECTION VII.
CONSTANT DESIRE TO MAKE WATER.

This is generally most troublesome in the latter months of pregnancy, but it is sometimes during the first months. It is owing to an iritable state of the womb after conception.

TREATMENT.--Light and cooling purgatives, such as antiquilious pills, oil, senna, rheubarb, cream of tartar. &c.

salts should never be taken in this complaint, as it has a tendency to increase the iratability of the parts; salts is by no means a good purge for pregnant females, it increases the excitement of the parts and leaves the bowels in a costive state. Warm clysters of slippery-elm tea, or new milk and water thrown up the fundament three or four times a day, will give great relief, and sitting over a pot or tub of water, every time she wants to make water, will enable her to pass off the urine with greater ease, it will also lessen the inflamatory condition of the womb. By pursuing the above course and avoiding violent exercise, this troublesome complaint may be mitigated and sometimes entirely relieved.

SECTION VIII.
STOPPAGE OR SUPPRESSION OF URINE.

Stoppage of the urine is not uncommon in the latter months of pregnancy. It is occasioned by the weight of the womb pressing on the neck of the water-bladder; this pressure prevents the water from flowing from the bladder at those periods when nature requires the evacuation.

TREATMENT.--The contents of the bowels must be evacuated by means of warm injections of milk and water, or slippery-elm tea. Apply cloths wrung out of warm water, to the lowest part of the abdomen or belly. Let the woman stand upon her feet and support the weight of the child with her hands, and endeavor to raise or change the position of the child; this often gives speedy relief. But if all the above remedies should fail, resort must be had immediately to a cathetar. Instruments of this kind may generally be had at any of the doctor shops and not unfrequently at the stores. For a description of the cathetar and the mode of using it, look under that head.

SECTION IX.
FLOODING.

Flooding, when in a state of pregnancy, may always be considered as dangerous, and requiring the immediate aid of a skillful physician. No discharge of blood ever takes place from the womb in a sound state of pregnancy, whenever such a discharge does take place it is proof that there is something wrong.

TREATMENT.--As soon as this dangerous complaint is discovered, put the woman in bed, keep her as cool as possible, and admit plenty of fresh air, give her nothing of a heating nature either to eat or drink. Give her a tea of *Oo-na-stah-lah-cah-stee-le-skee* to drink freely, if this cannot be had, turn to Materia Medica, there you will find many valuable astringent tonics fully described. Give any article of this class freely, without the apprehension of dangar from their use.

SECTION X.
FALSE PAINS.

Women are frequently troubled with pains which resemble labor in so many particulars as to give great uneasiness. False pains may be produced by costiveness, eating such articles of food as produce wind in the bowels, by fatigue, dysentary, &c. They may be relieved by regulating the diet, attending to the bowels, avoiding exposure and fatigue, and drinking freely of a tea of moccasin flower root, and the leaves of the English raspberry.

SECTION XI.
ABORTION.

Abortion may take place at any time during gestation after the first month, but it frequently occurs between the eighth and twentieth week. After a woman has once miscarried, she is much more liable to the same accident than before, and when a habit of miscarrying is once formed, it is a difficult matter to prevent it. Abortion is generally caused by sudden frights, violent fits of passion, external injuries, such as falls, blows, &e., violent puking or purging, oppressive exercise, excessive venery, or great uneasiness of mind. Abortion is generally preceeded by pains in the back, loins, and lower part of the abdomen, there will be a looseness or softness of the breasts, and a chilliness of the body. Slight discharges of blood will take place from the womb, which sometimes increases until it amounts to flooding.

TREATMENT.--When symptoms of Abortion appear without a belief that the child is dead, every possible means must be used to prevent the loss of the child. If she has fever and is of full habit, blood should be drawn, but if she be in a cold state which is generally the case, give stimulating and sweating teas. Let cayenne pepper be used freely, ginger and whiskey stew is an excellent article to warm the system. I forgot to tell you when speaking of flooding, that if the patient be in a cold state, warming teas, such as red pepper, ginger with spirits in it will be of great utility; but if the patient has fever and is flooding, cold applications should be used in their stead. The bowels should be well attended to and costiveness avoided by the use of injections, and mild purgatives. A tea of common hemp-seed may be used with great advantage, this tea alone often prevents abortion. Women who are liable to have discharges of blood from the womb during pregnancy, should make constant use of spikenard and columbo-root bitters. After symptoms of miscarriage, the patient should remain in bed for several days, and as she values her own safety and the life of her child she is to avoid all the above named causes which tend to produce accidents of this kind.

CHAPTER III.

SECTION I.
LABOUR

Labor means the interval of time between the period when the woman begins to be delivered of her child, and her final delivery. Although Labour is an operation purely natural, it is preceded by various symptoms which indicate its approach. A few days previous to delivery there is a shrinking of the waist, sometimes this does not take place until within a few hours of actual labor. Pains are next felt in the back, loins, and a slimy matter is discharged from the birth place, generally colored with blood.--The pains are at first short, and only return after considerable intervals; but they gradually increase in length and severity, and the intervals of ease are much shorter. There is often chillness, sickness and vomiting. To ascertain whether the woman is in actual Labour, the midwife is to introduce her finger to the mouth of the womb, (having first oiled it well,) if there is much pressure from above on the mouth of the womb, and if it appears to dilate or open during the continuance of a pain, the woman is in actual Labour. During the first stages of Labor, nothing is to be done by the midwife only to keep the mouth of the womb in its proper place; it is often turned so far back as to produce much unnecessary suffering, this is to be done with great tenderness on the part of the midwife. The contents of the bowels should be evacuated by injections or a dose of castor oil, the urine should be passed off whenever she feels the least desire to evacuate it. During the stage of Labor the woman is to be kept quiet, and all rational means employed to inspire courage, bear up the sinking spirits and prevent entire despondency. As the womb gradually becomes more and more dilated and the pains become more frequent and severe, the patient often becomes impatient and dissatisfied with every body about her, frequently demanding help from the midwife. This is a critical time and any force on the part of the midwife may do serious injury. She may lie in bed with a pillow placed between her knees so as to keep them sufficiently wide for the child to pass. When the pains become severe and bear down considerably, if she should desire to be placed in another position, her request should certainly be granted. Most women prefer sitting on the knees of another and this is certainly the most natural and easy position. As the pains increase the child's head descends rapidly at every bearing down pain, and soon fills the basin or pelvis--this is called the second stage of Labour, and it is at this time that great support must be given by the midwife in pushing with the palm, or soft part of the hand against the *perenium,* which is that part between the birth-place and the fundament. This must be strictly attended to, for it is easily torn, and when torn it can never be remedied. The lacerating or tearing of this part connects the opening of the fundament and birth place, and leaves the poor innocent sufferer in a most unpleasant and miserable condition through life. The third stage of Labour commences at the time the child's head starts through the external part of the birth-place. In this stage you are to be very particular in supporting the *perenium.* When the pains come on with violence and rapidity you are to push gently against the perenium in a manner, rather to retard than to hasten Labour, especially if the child's head seems to advance too fast. If it should advance very slow at this stage, the midwife may assist the birth by taking hold of each side of the head with her hands when it is sufficiently advanced, and pulling gently during the continuance of a pain. When the head is born, the mother will generally have a little rest, which should be alowed her--she should be soothed and cheered by the midwife. The hand must still be pressed gently on the perenium, bearing it somewhat upwards, this pressure must be continued until the hips and thighs have passed the mouth of the birth-place. When the child is born, let it and the mother be perfectly quiet for a few minutes, taking the naval-string between the finger and thumb, and so soon as the pulsation in the cord has ceased, tie a string firmly about three inches from the naval, then apply another tie about two inches from the first, still nearer the placenta or after-birth, then cut the cord between the ties. The naval-cord must in no instance be cut until the breathing of the child is established. Generally the child cries immediately after it is born, but if it does not, its mouth should be cleared of every thing that is calculated to obstruct breathing.--Its body should be wet with spirits, and the naval cord stript between the thumb and finger from the mother to the child. If these means fail, so soon as the placenta or after-birth is expelled, place the child in luke warm water and give an injection, in which put a portion of spirits. If all the above means should fail, take the child to a door or window immersed in warm water to the chin, and place the after-birth on a shovel of hot embers, stripping the cord from the after-birth to the child as above directed. When the child's head is born, the midwife should ascertain whether the naval cord is drawn about the neck of the child. If she finds it drawn around the neck of the child, she must gently draw it over from the back of the head to the face.

After the woman is delivered of her child, the placenta or after-birth has yet to come away. This commonly takes place in from five to forty-five minutes after the birth of the child. But, if nature should not expel the after-birth in the course of fifteeen minutes, the operator should move the cord very gently, but do not pull it, the patient may blow gently in her hands and rub the abdomen to aid the contraction of the womb. If these means should not cause the womb to contract and expel the after-burthen in the course of an hour, you may take the cord in the left hand, and follow the cord with the fourfinger of the right hand up the birth-place, and if you readily feel the root of the cord, continue rubbing the lower part of the belly with the hands, and let her continue blowing in her hands, these means will generally throw out the after-birth. But, if you cannot reach the after-birth with your finger, and nature seems quiet on the subject, a further examination is necessary. If the after-birth seems fast to the womb, take a part of it softly into the fingers and press it gently, still using the above mild means. If this should fail to expel the after-birth, you may feel cautiously, and separate between the edges of the after-birth and the womb, any parts which may adhere as the womb gradually closes. When the after-birth is expelled and any great discharge ensues, treat if as directed under the head *flooding*. But let me here tell you that more dangerous floodings are produced by hastening the expulsion of the after-burthen than in any other way. Many unskillful midwifes think the sooner they can expel the after-birth, after the birth of the child, the better. This is a very erroneous idea. and has caused the death of many a woman. As soon as convenient after delivery, the woman is to be placed in a clean dry bed, she may take some nourishment such as a cup of tea, coffee, or light penada. She should not encourage conversation, but remain quiet, and if she feels disposed to sleep, she should indulge it; her room should be kept comfortable, if the weather be warm, give her plenty fresh air, if cold, make it comfortably warm, and do not annoy her with company.

As soon as convenient, a broad bandage is to be placed round the abdomen, comfortably tight. This bandage is to be worn at least one month. This will prevent the woman from having an ill-shaped abdomen, after recovery.--The day following her delivery, she should take some mild purge, such as oil, cream of Tartar or rhubarb. If she is allowed to become costive, child bed fever may be expected. After having cut the naval-cord as above directed, you are to wash it clean with warm water and soap, and wipe it dry. The naval is to be dressed by burning a hole through a fine cloth of several folds, greasing the under side with tallow or oil, then drawing the naval cord through the hole, then apply a bandage round the belly comfortably tight. In four or five days, the cord will slough off; the naval should then be anointed with an ointment made by stewing heart-leaf root, or bearsfoot root in fresh butter. This ointment will soon remove the tenderness.

SECTION II.
UNNATURAL PRESENTATIONS.

A natural presentation is when the crown of the head presents, and the body follows in a straight line. When any other part of the child presents, it is unnatural, and will produce difficult labor. when the membrane containing the water has broke the midwife can easily ascertain what part of the child presents.

When the feet present she should endeavor to get them both, and the labor may be suffered to progress in the natural way: the midwife may assist the birth by gently drawing the child during each pain. She should with both her hands, bring down the arms along with the child.--When the breast and arms are born so far as the shoulders if the face of the child be not downwards, it must be turned to that position, in order to prevent its being stopped by the chin over the share bone; having brought it to the shoulders, the operation is to pull the child forward during the next pain; so that the head may take the place of shoulders, and not be stopped in the passage. The operator must be sure, that she has the feet of but one child, if there be two children, and she should get the left foot of one, and the right foot of the other, it would prove fatal both to the mother and children. To ascertain this, she is to slide her hand gently up the leg and thigh, until she finds them both joined to one body. When the breech presents, you should endeavor to bring it feet foremost as above directed. When the hands and feet present together, the hips of the woman should be raised a little higher than the head and shoulders; and when the mouth of the womb is sufficiently dilated, the hand is to be introduced far enough to reach the breast of the child, which is to be gently thrown back towards the bottom of the womb, leaving the feet in in the passage, after which the child is to be brought forth feet foremost as before directed. When the hand and shoulder presents, the operator should gently try to push back the part and keep it so, and in most instances if properly done, the pains will force the head into the pelvis, and bring the presentation to a natural one. But if these means should fail, the feet are to be searched for with great tenderness and caution. Having found and secured both feet, they are to be brought down by the child's face, for if brought down by its back, it will endanger the lives of both mother and child. If the woman is feeble and much exhausted, the delivery must be aided and hastened by the midwife.

SECTION III.
TWINS.

Twins are generally smaller than other children; and on this account, their birth is more rapid, and easy than that of single children. After the birth of the first child, it may be easily ascertained whether there is another child contained in the womb. When there is but one child the womb diminishes very much in size soon after the birth of the child, and the bowels which have been kept out of their natural situation, during the latter months of pregnancy, immediately get forward to the fore part of the belly, and render it soft and yielding. But where a second child remains, the womb does not appear to diminish in size, and the fore part of the belly has the same hardness as before delivery. Sometimes the pains advance rapidly and the infant comes soon after the first. In such cases, all the practitioner has to do, is to be assured that the child is in the proper position, and so conduct the extraction of the afterbirths, that no alarming discharges may follow.--When the pains cease, after the birth of the first child, the operator should not interfere, until the woman has measurably recovered from her fatigue. If the pains do not return for the expulsion of the second child, in the space of an hour, after the birth of the first child, give her a tea of red raspberry leaves, oneberry leaves, or white hazel leaves made pretty warm with red pepper. Rub the abdomen with the hand, and the womb will contract, and bring a return of the pains. The naval chord of the first is to be tied as directed for single children, securing the chord to prevent it from drawing back. The afterbirths, if there be two, will both be expelled at once, after the birth of the last child. When the first child presents natural, which is head foremost, the second may be expected to present feet foremost or in some worse position, it is to be treated as I have already directed for the different presentations.

CHAPTER IV.

SECTION I.
DISEASES CONSEQUENT ON DELIVERY.

AFTER PAINS.

These pains, are occasioned by the contraction of the womb, in its exertions to expel the clots of blood and secretions which are contained in the womb after the birth.--When they are not too severe, and produce but little inconvenience, it will be best to let them alone. But when they become so severe as to weary and weaken the woman, they are to be relieved by the use of a tea of red raspberry leaves and moccasin-flower root, in which put a tea-spoonful of the anti-spasmodic tincture to each half tea-cupful; the dose should be repeated as often as circumstances require it. This treatment will often relieve flooding after delivery.

SECTION II.
SOCIAL DISCHARGES.

The discharges, which take place from the womb and birth-place, for several days after delivery, are termed Lochia, which in English, means cleansing. If these discharges do not flow so plentifully as may be expected or if they entirely stop, no regard need be paid to this circumstance, if the patient be otherwise as well as can be wished, for this evacuation is not only different in different women, but even in the same women in different lyings-in, from which she recovers equally well. But if they should become scant or stop, and produce ill feelings, they are to be increased by the use of warm stimulating teas, and sitting over the bitter herbs. The birth-place is to be washed once or twice a day with warm milk and water, and occasionally thrown up the birth-place. This will greatly facilitate her recovery, and in many instances it will prevent disease.

INFLAMED OR SWELLED BREASTS.

Inflamed and sore breasts are caused by cold settling in them and obstructing the passage of the milk.

Let the patient keep the bowels regular by the use of mild purgatives, minding not to use salts. She should drink some sweating teas to keep up a perspiration, and bathe the breast frequently in a strong decoction of beach leaves or bark, and apply a poultice made by thikening the above decoction with wheat brand. The face of the poultice should be smeared with fresh butter, hog's lard or some kind of oil, to prevent its sticking. A beeswax plaster made as directed for mumps, and applied to the breasts, is an excellent remedy. If you find the above means will not prevent it from coming to a head, boil sumac-root bark in sweet milk, thicken it with flour and it will draw it to a head as speedily as necessary: when it is perfectly ripe, and not before, have it opened, and continue the application of the poultice.

SECTION III.
SORE NIPPLES.

Take red-oak bark and boil it until you have a strong decoction, then strain and continue boiling until it is reduced to the consistency of thick molasses; apply this extract over the nipple in the form of a plaster, and it will cure almost any sore nipple.

SECTION IV.
MILK-FEVER.

In a few days after delivery, the breasts become distended with milk. There is generally headache, thirst, hot, dry skin, quick pulse, &c. These feverish symptoms are occasioned by the change of the system after delivery, by the swelling and iritation of the breasts from the milk secreted in them. This fever may be relieved by taking a dose of cream of tartar to evacuate the contents of the bowels and cool the system, bathe the breasts with warm water or a tea of catnip. Drink mild teas, such as balm, sage, or hysop, and let the child suck frequently. Putting the child to the breast soon after it is born, and continuing to do so frequently, minding not to feed it, so as to prevent it from being desirous for the breast, will often prevent this fever entirely.

SECTION V.
SWELLED LEG.

This disorder may attack at any time, from the second to the fourth week after delivery. It is generally supposed to arise from some irritating matter being left in the womb. It is a complaint that seldom occurs where due caution and cleanliness are observed. The symptoms, dejected spirits, general uneasiness over the system; these are succeeded by a pain inside of the leg, extending from the heel to the groin--the slightest motion gives great pain. There is a slight pain about the womb, the discharges from the birth-place become very offensive. The pulse becomes quick, the skin hot, the tongue white, the urine thick. In a short time, the leg begins to swell and the skin turns of a pale glossy color, its peculiar appearance has given it the name of the White-Leg.

TREATMENT.--When symptoms of this complaint appear, lose no time in injecting forcibly up the birth-place warm milk and water, for the purpose of cleansing the womb of any irritating matter that may be in it. If she be costive, let her use cream of tartar freely, or in sufficient quantities to regulate the bowels. Steam the leg over bitter herbs or bathe it in a strong tea of catnip, after which, bathe it with a decoction of pepper and vinegar, and keep it wrapped in flannel. This course is to be pursued until relief is obtained.

SECTION VI.
CHILD-BED-FEVER.

This disease is technically termed *puerperal fever,* and it may be regarded as one of the most fatal diseases to which lying-in women are subject. It usually attacks in a few days after delivery, when it does occur. The symptoms are, chilliness, soreness about the womb, fever soon follows, and generally ceases in a profuse sweat, the sweat soon dries up and the skin becomes dry and burning, the face is flushed the thirst great and the tongue whitish. There is great pain in the head & back, sickness at the stomach, and sometimes vomiting. In a short time the belly swells, feels full and becomes very painful, so much so, that the lightest covering cannot be borne without giving pain. In some cases the bowels are quite loose, and in others much constipated or bound. The social discharges cease, the milk dries up and the breasts become flabby.--When this fever continues for a time, it is very apt to change to a typhus fever. This is marked by an abatement of the inflamatory symptoms, the tongue and teeth now become covered with a dark brown coat, small sores break out in the mouth and throat the breath smells badly, the stools are dark and very offensive.

When *Child-Bed Fever* changes to the *Typhus,* refer to the head of Nervous Fever, where you find the proper treatment. The treatment in the first stage, or in *Child-Bed Fever* properly so called, is as follows:

TREATMENT,--During the cold stage, warm applications to the feet will be of service, she may also drink warm teas, such as balm, sage, &c. When the hot stage comes on she is to be bled from the arm. After being bled she is to take a purge of cream of tartar, rheubarb, senna, or anti-bilious pills. If there is sickness at the stomach it will be proper to empty the stomach with some mild emetic.--Apply a cloth wrung out of hot water to the belly, and inject warm milk and water up the birth place this will lessen the pains and inflamation, cloths wrung out of a strong tea of catnip is an excellent application over the belly. When the fever is on give ipecac every hour in small portions, so as to produce slight sickness at the stomach, and gentle moisture of the skin. Injections of slippery elm tea will he of service. After the stomach and bowels have been attended to as above directed, give a sweat of senaka snake root tea, after which let the patient use mild teas to keep up a gentle perspiration. At the commencement of this disease the diet must be very light and cooling but as the disease advances, and she becomes weaker the nourishment should be increased.

The spirits of turpentine taken in table-spoonful doses every morning, in a mucillage of peach tree gum, and followed by a table-spoonful of castor oil in the evening until the violence of these symptoms cease, is said to cure this dangerous disease, with much more certainty than blood-letting. When the loss of strength is great, from purging, or from other causes or if the disease seems to be approaching the *typhus or nervous fever* it will be necessary to support her system, by the use of good wine or toddy-Equal quantities of wild cherry-tree bark, dog-wood bark and swamp-poplar bark, boiled together and the decoction perfectly cooled, taken in doses of a wine-glassful three or four times a day, given when the patient is clear of fever and weak, will be found an excellent strengthener of the system.

CHAPTER V.
FALLING OF THE WOMB.

This disease is common both to the pregnant and unpregnated state. It is brought on by going about too soon after delivery or the monthly discharge, before the womb has gained its usual tone and strength. It may also be brought on by jumping, or some violent exertion during or soon after labor. The symptoms are: a sense of bearing down pains in the back, groins and privates, and if the complaint is suffered to progress, the urine is frequently stopped by the womb descending into the vagina and pressing on the neck of the bladder. In the worst stage of this disease, the womb protrudes beyond the mouth of the vagina a considerable distance.

TREATMENT.--In the first stage of this dssease, or before it has protruded without the vagina, it may be relieved by bathing the small of the back and lower part of the abdomen in the essence of red-pepper. (See Dispensatory.) and applying a strengthening plaster to the back; she should bathe every twelve hours. To restore the general health, let her drink bitters of white sarsaparilla and wild mercury in good whiskey. When it has protruded without the vagina and made its appearance, take five pounds of white-oak bark, boil it in two gallons of water, down to three pints, then strain the decoction and add three pints of honey, mix it well and simmer it a few moments over a slow fire. The womb must be washed with casteel soap, and then bathed in strong alum water, and then anointed with the white-oak syrup and honey. It is to be dressed in the above manner every twelve hours. After it has been washed and dressed as above directed it is to be supported by a bandage and cushion. There is to be a bandage worn around the body just above the hips, and the bandage which is to confine the cushion to the womb, is to be fastened to this belt or bandage before and behind. The cushion should be covered with a fine, soft cloth every time the womb is thus dressed. She is to drink the same bitters as directed in the first stage, and bathe the abdomen and back in the same manner, and also applying the strengthening plaster to the back. This course is to be regularly pursued until relief is obtained. The above remedies effected an entire cure where the womb had been protruded without the vagina seven years and the woman was rendered unable to go about.--The womb returned to its proper place and she recovered her health in a few months. The patient should confine herself as much as possible to lying and sitting; she should avoid violent exercise, and, above all, lifting and stooping.

CHAPTER VI.
DISEASES OF CHILDREN.

Many of the diseases that children in common with grown persons are subject to, I have described, and the general treatment laid down in the foregoing part of this work, it will therefore be unnecessary for me to enter into a minute detail of all the diseases to which childhood is exposed.--I will confine myself principally to the diseases of infants, and such complaints, among children as have not been previously treated on in this work.

SECTION I.
RETENTION OF THE MECONIUM.

All children at their birth have a dark colored matter in their bowels called by phisicians Meconium. If this matter be not discharged in a reasonable time, it produces irritation and disease. Nature seems to have designed that the first milk drawn from the mother's breast, should so operate upon the bowels of the infant, as to evacuate the offensive matter by stool. This is amply sufficient reason, I think, why infants should be put to the mother's breast as early as possible after birth. If the milk should fail to operate on the bowels, it will be necessary to give some mild cathartic, sometimes a little molasses and water will answer; it should be given at intervals, until it operates. A tea of rose leaves is very good, or senna tea may be given: but if the above simple remedies should fail, give castor oil, or rhubarb; either of these will remove the meconium. Injections of cow's milk sweetened with molasses or sugar, will greatly aid the other remedies in cleansing the bowels.

SECTION II.
RETENTION OF THE URINE.
(Oo-ne-te-skay-ne-tsa-yoh-oo-lah.)

New-born infants are not unfrequently afflicted with suppression of urine, and if the difficulty be not removed, it will in a short time produce great pain, and in many instances, convulsions which sometimes terminate fatally.-- Particular attention should be paid by the nurse, to see if the child discharges urine, in a reasonable time, and quantity, after its birth, and also that it continue to do so at proper intervals afterwards.

TREATMENT.--Give a tea of pumpkin seeds, or water-melon seeds sweetened with sugar or molasses; it should be repeated frequently until relief is obtained. The abdomen should be bathed with warm water, and gently rubbed, a little spirits in which camphor has been dissolved, may be added to the water intended for the bath, with the happiest effects. Many valuable articles for suppression of the urine, is named and fully described in the materia medica, in the class of diuretics, and by turning to that part of this work, you will doubtless be enabled to administer relief to the little sufferer without delay.

SECTION III.
SNUFFLES, OR STOPPED NOSE.

This stoppage of the nose is quite common among young children, and is occasioned by cold.

Open the bowels with castor oil, or some other mild purge, grease the forehead, and across the nose with tallow or sweet oil, or bathe it with warm vinegar and water. If this does not give relief in addition to the above remedies, bathe it in warm water, and after the bowels have been opened, give it a little finely powdered sulphur, in some sweating tea, such as ground ivy, penny's worth or something similar. By this means, relief will be obtained, and in many instances, a severe attack of croup prevented.

SECTION IV.
SORE EYES.--(Tsi-loh-nah-ka-wis-kah.)

Keep the bowels open with rose-leaf tea, peach-flower tea, or any mild laxative, and wash the eyes with warm breast-milk, or water in which a little borax has been dissolved. Sore eyes are mostly caused by cold or derangement of the bowels, and by attending to the above prescription, and keeping the child from taking cold, there will be little or no trouble with the eyes.--Water, in which has been soaked the inner bark of sassafras or slippery-elm forms a good wash for sore eyes.

SECTION V.
RED-GUM.

In this disease small red or yellow pimples break out on the face, neck, and often over the body, containing a watery fluid. The child appears sore, and frequently screams as if pins were sticking in it, when it is handled.

TREATMENT.--No outward application should be made to remove this eruption, for by so doing, you might suddenly drive it from the surface, that is, strike it in, and thereby destroy the life of the infant. The child, while afflicted with this complaint should be kept from fresh, damp or cold air. The bowels should be kept open by the use of castor oil, senna or rhubarb, and from half to a whole tea-spoonful of sulpher should be taken in sage or ground ivy tea The only danger in this disease is in driving it in, when it is driven in by cold, it produces great restlessness and misery and not unfrequently convulsions. Should the eruptions suddenly disappear and the child become sick, put it immediately into warm water to the neck and give it some sweating tea, with a little sulpher in it. This should be repeated every hour or two until the eruptions again appear and a moisture is produced.

SECTION VI.
YELLOW-GUM.

This complaint in infants is similar to jaundice in adults. The skin becomes yellow, the eyes and urine are tinged with yellow, and the stools are clay-colored, and the bowels generally costive; in some infants this complaint takes place in a few days after their birth.

TREATMENT.--Give a purge--rhuebarb, I think, is some the best, but the butternut and gulver syrup will answer. If these cannot conveniently be had, give castor oil, senna, or peach flower tea, until the bowels are cleansed. After the bowels have been cleansed, give a strong tea of wild cherry-tree bark or poplar bark, in tea-spoonful doses three or four times a day, still keeping the bowels open by the use of cathartics until a cure is effected. If the child should appear sick, it would be well to bathe it in warm water several times a day, until it appears relieved:

SECTION VII.
THRUSH.--(Oo-hah-lah-go-huh-skee.)

This is a very common complaint among infants or small children, and is caused by a foul stomach and bowels. It sometimes appears in small white specks resembling coagulated milk on the tongue, inside of the lips and corners of the mouth. At other times it makes its appearance in yellow blisters on the tongue, gums and inside of the mouth. In some instances when suffered to run on a long time, it extends down through the alimentary canal and shows itself at the fundament. Some children afflicted with Thrush are very costive, while others suffer greatly from bowel complaint.

TREATMENT.--If the bowels are costive, the first step should be to evacuate their contents by the use of cathartics, such as rhuebarb, senna, castor oil, or butter nut and gulver syrup. If the bowels are distressed or laxative, give a tea of sweet-gum bark. The mouth should be frequently washed with water in which borax has been dissolved, solved, and the child permitted to swallow some of the same; borax finely pulverized and put into the mouth and let it dissolve by degrees, is a very good mode of applying it. A mouth water made of persimmon bark and dirt out of a chimney, where fire has been kept for some length of time is very good; the bark and dirt should be boiled, the decoction strained and sweetened to a syrup. Hens oil will often cure Thrush, the mouth and gums being frequently greased with it. The child should always swallow a portion of the mouth water, as the stomach is always more or less affected. Any stringent tea will be found good for Thrush, and by turning to the Index, you will be referred to various kinds of this class of astringents. In the Dispeusatory you will find several valuable recipes for mouth-water, used in the cure of this disease.

SECTION IX.
CHOLIC IN INFANTS.--[Tsu-ne-yah-low-tis-scoh.]

This is a very common and troublesome complaint among young children and is easily known by the child suddenly screaming and crying, at the same time drawing up its legs; the complaint is sometimes so severe that the child cannot make water. This complaint is not unfrequently produced by over-feeding and suckling small children, by costive bowels, and by taking them out of warm rooms into cold or fresh air, or by putting damp or cold clothes on them.

TREATMENT.--If cholic arises from flatulence, give a tea of peppermint, ginseng, angelica, calimus, or all-spice, to which you may add a few drops of paragoric, Bateman's drops, or laudanum; if the laudanum is used, do not give more than from one two drops. The essence of peppermint or pennyroyal will be found most valuable remedies in flatulent Cholic among infants. Bathe the infant's belly before the fire, and rub it with spirits in which camphor has been dissolved, a few drops of laudanum may be added to the bath, if the child appears very much distressed.

When cholic arises from acidity, it may be known by the bowels not being bound and the stools of a green color and sour smell. In addition to the above remedies, give the child, occasionally, a dose of magnesia or the infusion of rheubarb in small doses to regulate the stomach and bowels. However easily the Cholic may be relieved very desirable that we do something to prevent its attacks, for let the cure be as speedy and as easy as it may, the preventative is always preferable. To prevent cholic in infants, let the mother drink daily of spirits, into which has been put asafoetida or garlic, and some of the same may be given to the infant occasionally; the chidren of women who make constant use of either of these articles while suckling, have very little if any use for laudanum, paragoric, Bateman's drops, &c., &c.

SECTION X.
CROUP.

Children of both sexes and all countries, from infancy, up to ten or twelve years, and even older, are liable to this complaint. It is a very dangerous disease, and the rapidity with which it proceeds, requires immediate and prompt attention. The head of every family should be acquainted with the symptoms and treatment of Croup, for of all diseases to which children are liable, it is certainly the most dangerous, and in many instances it makes its advances with such amazing rapidity that suffocation takes place before medical aid can be procured. The most fat, gross children are most liable to it, and next to them, the extremely puny. It often seems to run in families, and is thought by some to be hereditary.

SYMPTOMS.--It very often attacks suddenly, with a short dry cough, hoarseness, wheezing, and great difficulty of breathing. The face looks flushed and swelled, the child throws its head back, its mouth open, and its eyes and neck veins look like they would burst. Sometimes a cold, clammy sweat breaks out. At other times this complaint comes on very slow, either with hoarseness and symptoms of a common cold, or else with looseness of the bowels, as if from cold, every symptom gradually becomes more and more violent, until a spasm attacks.

TREATMENT.--On the first appearance of Croup, give onion syrup, and apply a poultice of onions or garlic to the throat and breast. It the bowels are costive, lose no time in evacuating their contents by injections of warm milk and water, with a little lard and salt in it, repeat the injection until the bowels are cleansed. If the attack be violent, put the child in warm water up to the chin, give it a puke of the tincture of lobelia and Indian physic, if the

puke should be slow in operating, wet tobacco leaves in warm water and vinegar, and apply it over the breast and throat. If the child is so far spent as to be unable to swallow, the emetic or puke may be given by injection, at the same time, applying the tobacco externally as above directed. In children of full habit, where the attack is violent, blood should be taken, either from a vein or by cupping; for several days after the spasm, the bowels should be kept open by the use of sulpher taken in some sweating tea, such as spice-wood, pennyroyal, &c. In nine cases out of ten, a strong tea of the fine roots of spice-wood, made very sweet and combined with sulpher in tolerable large doses, will entirely relieve croup if given in the first stages. We copy the following from the writings of a respectable physician among the whites: "After using the remedies generally prescribed without success and the case is desperate, the best remedy I have ever tried as a last resort, is calomel in large doses, from thirty to fifty grains, repeated in ten grain portions every fifteen minutes, until relief is obtained, (do not be alarmed at this dose,) I know by experience, in a hundred instances, of the lives of children being preserved by large doses of calomel, which otherwise must have been lost. Then let me urge upon you the necessity of laying aside your prejudices against this medicine, and not to slacken your hand in this trying moment if you wish to preserve the infant. So powerful and salutary is this medicine, that it frequently relieves the complaint in ten or fifteen minutes, without recourse to any other means." I give the above in order that those who may choose the use of calomel in cases of emergency, may give it in doses sufficiently large to produce the desired effect.

SECTION XI.
WORMS.

There are a variety of Worms which infest the human system, and grown persons are often troubled with them as well as children. But as it is generally viewed as a childs complaint, I have thought proper to give it a place among the diseases of children, but will try to give a satisfactory treatise on the subject.

The first class are those which inhabit the whole range of the intestinal canal, they are the long tape-worm, the broad tape-worm, the flat two-headed worm, the long round worm, and the long thread worm. Those which inhabit the lower part of the intestinal canal, belong to the second class, as the thread-worm, the hot-worm and the mane-worm. There is another or third class, as the maggot-worm, &c. Worms are sometimes passed with the urine, and must originate in the bladder and kidneys.

SYMPTOMS.--The symptoms indicating worms are various and contradictory, often resembling the symptoms of other diseases in so many particulars, as to render it difficult to determine whether worms really exist. The child or person, in general, has a larger belly than usual, and pains are felt in the stomach and belly; the appetite is various, constant hunger, and yet the system becomes weak. At times, there is sickness at the stomach, and vomiting; looseness at the bowels, interrupted sleep, bad breath, picking at the nose, a peculiar paleness about the mouth; a short dry cough, slow fever, and sometimes convulsions.

TREATMENT.--The roots, leaves or seeds of the Jerusalem oak, boiled in sweet milk, the decoction made sweet with honey given in such portions as the stomach will bear, of a morning, on a fasting stomach, and repeated at night, will carry off worms in a surprising manner; it should be repeated for several days, still giving a purge of butternut and gulver syrup; every other day cleanse the bowels.--Wild ginger boiled in sweet milk, and given in the same way, is one of the best warm medicines in the world.--Spirits of turpentine is a valuable article for worms; to children, give from a half to a whole teaspoonful in sugar, and repeat every morning for three mornings, and the third day give a purge, and if the patient is not entirely relieved, repeat the course in a few days. Carolina pink root given in powder or decoction, is good to expel worms: it should be given in honey or molasses; when it produces any unfavorable symptoms, give a purge, and all will be well. The China tree is a valuable warm medicine; the fruit, bark of the root or bark of the tree may be used: when it produces giddiness in the head, give a purge.--The yellow poplar root bark in powders mixed with sweetning, is good for worms, and as a family medicine is not surpassed by any thing in my knowledge as a preventative of worms, for puny children and such as are frequently troubled with warm spells; it acts like a charm; it should be taken every morning in honey, sugar or molasses; a tea spoonful is a dose. Those who wish to avoid worm sickness among their children, need only to give the poplar root every morning on an empty stomach. The China berries mashed, and put in spirits, is also very good to prevent worms. When persons commence doctoring for worms, they should continue it for several days and weeks, unless relief is obtained; but after a fair trial has been made, without its having produced the desired effect, there are strong grounds to suppose, that some other disease, and not worms is the cause of bad breath, for in many instances, children have been physicked to death for worms, when they had none, but were laboring under some other disease. In the class of Anthedmetics, in materia medica, you will find many other remedies for worms. When worms produce choaking, give honey and the patient will experience considerable relief, then give warm medicine to expel the worms, and the cure is effected. When taking medicines for worms, the patient may eat as much honey, sugar or molasses as he desires.

SECTION XII.
CHOLERA INFANTUM OR PUKING AND PURGING OF INFANTS.
(Tah-lo-ne-tse-tah-ne-gah-slee.)

This disease in infants is the same as cholera morbus in grown persons. The symptoms are, vomiting and purging, a quick pulse, hot skin, and great and constant thirst. It makes its appearance in the heat of summer, and its severity is in proportion to the heat of the weather, it generally prevails as an epidemic when it takes a start in a section of country.

TREATMENT.--Give a tea of cholera-morbus root, until the vomiting is checked, next cleanse the stomach and bowels with gulver syrup. After the first attack is relieved, give the child a tonic or strengthening medicines.--Yellow poplar bark will be very good, or the little white root called by Cherokees cul-sa-tse you-stee will be found excellent.

INJECTIONS OR CLYSTERS.

When giving the treatment of the different diseases. I have frequently directed the use of injections. It will probably not be amiss to give some farther directions, as to the mode of administering injections &c. When the injection is to be administered by an assistant, the patient is to be laid on a bed near the edge, the knees are to be drawn up, the attendant is then to take the clyster pipe, the finger is to be placed before it to keep in the contents, after it is well oiled it is to be placed to the fundament.--The pipe is to be gently pushed up the fundament about an inch, and the contents are to be forced out, by gently pushing the handle of the syringe with one hand, while it is firmly held by the other. Injections are sometimes given by putting the clyster into a bladder and injecting it up the fundament by means of a quill. After the quill is introduced up the fundament the bladder is to be squeezed, together with the hands.

Clystering is one of the most mild, innocent, powerful and safe remedies known in the science and practice of medicine.

Any medicine given by Injection should be given in much larger portions than when taken by the mouth.

PART THREE.

INDIAN MATERIA MEDICA.
PRINCIPALLY BOTANICAL.

Materia Medica means nothing more than simply the materials of medicine; it is that part of the medical science which treats of the nature, and properties of substances, whether simple or compound, mineral or vegetable, which are employed for the arrest or cure of disease, and for restoring health.

The Indians derive the materials of medicine employed by them in the healing art, almost entirely, from their own native forest. Notwithstanding their remedies may appear simple to those unacquainted with their medical properties; yet it is frankly acknowledged by the whites who have had an opportunity of personal observation, that in many instances they have arrested disease, with apparent ease, when the remedies prescribed by white physicians of character and skill had failed.

Botanical materia medica is yet in its infancy, but little has been done towards classing the articles composing this branch of medical science. We shall attempt to class them, according to their most active medical properties We deem the classification here proposed, best calculated to simplify the healing art, and thus render it more readily comprehended, and consequently more useful to the community in general. The articles described under one class, in many instances, possess also the specified properties of other classes, though in a less active degree. In such cases, I will place the article under the head, where I think its most distinguishable, and active medical properties entitle it to be placed; but at the same place, I will describe all its medical qualities, so far as I have learned them.

The Indian name of the articles will be placed at the head, immediately under the Indian name will be seen the most common names applied to them by the whites, and to the right of the common name will be given the botanic technical name--so far as I am acquainted with them.

In some instances the opinion of white physicians with regard to the medical qualities of different herbs or roots are given, in such cases it is given as their views and plainly distinguished from the Indian theory.

CLASS No. 1.

EMETICS.

Emetics are a class of medecines, which on being received into the stomach, produce vomiting, or puking--they are called by physicians emetics, and are given in a great variety of cases to rid the stomach of its noxious contents. Their operation will always be increased, and rendered much easier, by drinking water made milk or blood warm, in considerable quantities after the first operation.

IPECACUANHA--(Raicilla.)

Ipecacuanha is a native product of South America, and this word in the Spanish language signifies emetic root.--This name is applied in Spanish America to various plants that possess emetic properties to any considerable extent--this fact gives rise to the confusion which is so common concerning this plant--it also accounts for the several varieties found in the shops, bearing the same name. The botanical name for this root is Raicilla. By paying attention to the proper or botanical name, the genuine article may generally be procured from the shops without difficulty. The genuine ipecacuanha, in its dry state, is a small wrinkled root, about the size of a hen's quill, variously twisted, and marked with projecting parts, apparently like rings--ash colored. It has a sickening and slightly bitter taste, and very little small. The outer bark is very brittle and it is, in this part, that the power and activity of this root as an emetic resides. This is generally sold in the shops, in the form of powder; that being the form in which it is administered as an emetic or puke. The powder is the color of common ashes. I have now described to you the imported ipecacuanha or the medicine which is most commonly used by physicians, among the whites under that name. I will now give for the further information of the reader, the opinion of Dr. Gunn respecting its medical properties. He says: "I may justly remark, that it stands at the head of the vegetable emetics for promptness, efficacy and safety of its operations. In powder, which is the manner in which it is generally given, full vomiting will be produced in a grown person, by a scruple or half a drachm, or you may put a tea spoonful of the powder, into six table spoonful of warm water, and give a table-spoonful every few minutes until it operates; or you may steep it in wine, and give it in small doses, until the effect you desire is produced.

The medical uses of this powder, when properly applied are very great and valuable. In addition to acting as an emetic, it will, when given in small doses, so as to produce nausea (sickness at the stomach) generally produce moisture of the skin or evacuation of the bowels, and in still smaller doses, generally stimulates the stomach, increases the appetite, and assists digestion. When given in small doses, it acts, not only as a diaphoretic, (sweating) but as an expectorant, (which means a free discharge of tough mucus and spittle from the mouth and throat.) It is also valuable, when given in small doses, to stop violent hemorrages (bleedings) from the lungs and womb. In intermittent fevers, it has generally succeeded in stopping them, especially when given about an hour before the coming of the fever, and also when given so as to produce vomiting at the time of the fever or end of the cold stage. Great benefit is often derived from the medicine, in continued fevers-and particularly in the commencement of typhus fever; an emetic or puke of ipecacuanha, followed by a sufficiency of this medicine, in very small doses, to keep up a gentle moisture or sweat, will, if attended to, in the early stage of this complaint, probably at once cut short or greatly lessen the severity of the attack."

I have given the above views of Dr. Gunn, for the information and benefit of those who prefer the imported to the American ipecacuanha.

In my practice, the American ipecacuanha or Indian physic is most generally used.

AMERICAN IPECACUANHA.
[UHE-LAY-OO-YAH-TER-TEE.]

This very useful, and somewhat singular plant, is said to be exclusively a native of the United States, and may be found in great abundance, in the middle, southern and western States, growing in loose, moist, sandy soils, and frequently in beds of almost pure sand. The leaves of this plant vary so much in shape and color, and in fact, the

whole plant varies so much in its different situations, that it is often mistaken by those unacquainted with its habits, for several distinct species of plants. It has a large, long, white, or yellowish colored, perrenial root, which sends off towards the upper part, many small roots about the size of small quills. The stems are numerous, of a reddish, pale green, or yellowish color. The leaves grow opposite to each other, and are generally of an oval form--but sometimes, they are of a long, oval--sometimes pointed--and at others linear. The flowers appear in the month of May, at which time, the leaves are very small, but as it advances in age, they become greatly increased in size.--The flowers are succeeded by three square or triangular capsules, or seed vessels, each capsule containing three seeds. The root of this plant is the part used for medicine, and is, in my opinion, far superior to the imported Ipecacuanha. It is a powerful emetic, both safe and certain in its operations, and is applicable to nearly all the cases, in which emetics are required. In doses of from five to ten, or fifteen grains, it is an excellent emetic: in doses of twenty grains, it operates as an active purge. In very large doses, it produces, in addition to the above effects, vertigo (giddiness in the head,) heat and great prostration of strength. When it is not convenient to weigh this article, put from one to one and a half tea-spoonfuls in a half pint of hot water, and when it becomes sufficiently cool, give it in table-spoonful doses, at intervals of from five to ten minutes, until vomiting is plentifully produced, aid it by the free use of warm water; after every motion to vomit, when the stomach is sufficiently cleansed, give gruel which will turn it to the bowels. This is an excellent article in Asthma, (phthisic,) colds, &c. Take in good spirits, a handful of the root to a quart of spirits, drink enough of this tincture twice a day, to excite slight (nausea) sickness at the stomach; it is also very good, taken in this way, for an inactive state of the liver--this tincture may be rendered better for the liver, by adding to the Ipecacuanha equal quantities of gulver root, and boneset leaves. For a full description of these herbs, refer to their heads. The American Ipecacuanha is much stronger than the imported Ipecacuanha. In administering the American Ipecacuanha for an emetic, I have added one fourth gulver, and found it much better than the Ipecacuanha used alone.

The root is the part used.

INDIAN PHYSIC.
[UH-LAH-OO-YAH-TER-TEH.]

Some writers among the whites make no distinctions between this plant and the American ipecacuanha, while others represent them as being two distinct species. It is not material which of these opinions be correct if we but understand the medical properties of both. We believe the Indian Physic to possess the same medical properties as the American Ipecacuanha and may be used to advantage in all cases where the Ipecacuanha would be advisable. It is combined with other articles in almost all cases where an emetic is required, by the Cherokee Indians. It is to be found in great abundance, in almost every part of the Western country, inhabiting shady woods and the sides of rich hills on mountains, from the Lakes of Canada to the Gulf of Mexico. It has a perennial roof, composed of several long, slender, brown-colored branches, issuing from one common caudex or head, and running some distance under ground. The stems which rise from the root, vary in number. Sometimes there is but a single one, and at others, there are many--they are round branches towards the top and commonly of a redish color and grow to the height of from two to three feet. The leaves are of a deep green color, long and pointed, flowers white. I again tell you that this root is valuable in all cases in which the American Ipecacuanha is recommended.--When administering it as an emetic, I combine it with gulver root, sometimes equal quantities of each, and sometimes two-thirds gulver to one of the Indian Physic. In preparing the emetic take the root after it is well sleansed and boil it in water to a strong decoction and give a half gill of the decoction every fifteen minutes until vomiting is produced. Encourage the operation by the free use of warm water; after the stomache is properl cleansed, give gruel to determineit to the bowels. The tincture of this root is an infallible remedy for that dreadful malady Milk-Sick, as may be seen in the treatment of that disease.--This root also forms an ingredient in Foreman's anti-bilious pills. Combined with bone-set leaves and gulver root, it forms a valuable bitter for an inactive state of the liver and spleen.

LOBELIA, EMETIC HERB, &c.--(Lobelia Inflata.)
[TOS-LAH-E-YOU-STEE.]

Lobelia Inflata is a beautiful plant, that is, it requires two years from the time it comes up from the seed before it comes to perfection and produces its seed. The stem is erect, milky, branched, growing from one to two and a half feet high. The leaves are alternate, oblong, acute, serrate and sessile, they contain like the stem a milky substance. The flowers are small, of a pale blue or whitish color, and generally put forth from the stalk solitary, immediately above each leaf. The seed vessels are small, oblong, roundish pods, seeds many, very small and of a brown color.

The first year this plant only sends forth a few radical, roundish leaves, which lay close to the ground--the second year it produses the branches and seeds. Lobelia is a common plant in many parts of the United States, growing mostly in old fields and open lands, rarely in the woods.--It is often found growing in great abundance in stubble fields, if left uncultivated the next year after the crop is taken off. When any part of this plant is broken, a milky substance or juice exudes from the wound, of a most penetrating, diffusible nature. This juice when applied to the eye has a most powerful effect, and hence it is called eye-bright; it has also received the name of Indian-tobacco, from the circumstances of its being used as a medicine, by them, and was thought, by some of the whites, to produce effects similar to the common tobacco.

Dr. Thompson, and several later writers on the same system, esteem it as being far superior, as an emetic, to any other article of that class in the compass of their knowledge--it is their alpha and omega in all cases. But we beg leave to differ with them on this subject, (as well as on many others,) we believe it to be two severe in its operations for weak breasted persons. It is an excellent medicine in case of croup and asthma or phthisic. The tincture given in small doses is good for infants, to prevent cholics, croup, &c., for this purpose it should be given in very small doses, just sufficient to produce slight nausea.

The most important use we have for this herb, however, is in the bite of reptiles and stings of insects. In the bite of the copper-head or rattle snake, we give it as an emetic; the tincture is preferred, but if this is not at hand, bruise the herb, put it in spirits, shake it well, and administer it in large drinks until copious vomiting is produced. If the other articles directed for external application, are not at hand, this tincture or bruised herb may be applied to the wound, until some of the other articles can be procured. The tincture forms an excellent application for the sting of insects or bite of spiders.

This plant may be used at any age from the time it first springs up until it gets its full growth and ripens--the same quantity possessing at all ages the same strength or virtue. The best time for gathering this plant is in the latter part of the second season, when it has arrived at maturity, which may be known by the leaves and pods beginning to turn yellow, at which time the seeds are ripe, and they are the most active part of the whole plant.--When it is gathered, it should be kept clean, and the night air excluded as much as possible. The tincture is made by putting the pulverized leaves or seeds into proof spirits and let it digest ten days in sun heat, the green leaves will answer. This article may also be used in an infusion in warm water, not hot, for anything above blood heat destroys its virtues and deprives it of its emetic property.--When you wish to give this herb as an emetic, in any case except snake bite, it is best to give it in small doses, repeated every few minutes until it operates; in snake bites it must be taken freely. The powdered leaves generally require a tea-spoonful and sometimes more to produce vomiting--of the tincture from one to two table-spoonsful. In case of snake-bite it requires much more.

VERVINE OR VERVANE--(Verbena Hastata.)
[TE-SCO-YUR-TE-NEEN-WO.]

This well known plant is commonly found growing in uncultivated fields, fence corners and about yards.

This herb forms a tolerable good emetic, but not equal in our estimation to the Indian Physic. For an emetic, it may be used in a decoction of the green or dry herb, or in powder like lobelia. The leaves, seed and roots, are all used for medicine, but I prefer the root in all cases, except for an emetic. The root combined with black-root and puccoon-root and digested in any kind of spirits, except peach brandy, and taken for chills and fevers, will be found of great benefit. A single handful of this root and a table-spoonful

of steel-dust, put into a quart of prooff spirits, is good for Dropsey. The decoction is also good for flux and old bowel complaints, either among children or among grown persons. The decoction given in early stages of fever seldom fails to throw it off entirely. It is an excellent sodorific and is valuable in decoction for colds, coughs, female obstructions and afterpains. It ha been employed with considerable success in the treatment of consumption.

INDIAN FEVER ROOT.

This plant has a perenial root, which runs horizontal and sends off many fibers--this root has a dark color and a strong rich taste. Several stems arise from one root to a height of two or three feet, they are round, faintly striped and covered with hair or down which is scarcely perceptible, the leaves are scattering, supported on long foot stalks, which sheath the stem, and like it are covered with down, which becomes hard and rough as the plant attains maturity. Flowers are large and yellow.

This root is much used in the cure of fevers by the Cherokee Indians, and it seldom fails, in their hands, to effect a speedy cure. A strong decoction of the root when drank freely, operates as an emetic and gentle purgative, and promotes a free perspiration. It is a very valuable article in colds and female obstructions. It may be used alone or combined with other emetics.

GILLENIA--(Gillenia Trifoliata.)

Gillenia grows in rich shady woods, where the soil is light, and has a branched and very knotty root, several, smooth, slender, erect stems spring up from the same root, considerably branched and of a reddish tinge, the leaves are alternate and slightly toothed, the flowers few and scattered.

A decoction made by boiling this root in water has a beautiful red color, and a very bitter taste. It is a valuable emetic and may be taken in decoction, or the powdered root taken in warm water, until vomiting is produced.--Dose for an adult is a tea-spoonful of the pulverized root put into a half gill of hot water, one-fourth of which must be taken every fifteen minutes until it operates.

CLASS NO. II.

CATHARTICS.

Cathartics are a class of medicines, which increase the frequency of the stools, by quickning the peristaltic motion. Of this class of medicines there are two kinds, which are distinguished according to their degrees of activity.--

Those which operate with mildness are called laxatives, and those which operate with violence and activity are called purgatives, the harshest of which are called drastic purges. Those articles both laxatives and purgatives, will be placed under the general head cathartics.

Very few diseases to which the human family are subject can be relieved without the use of medicines of this class, which is fully shown in the treatment of the different diseases.

BUTTERNUT, WHITE WALNUT.--(Inglans Cineria.)
[CO-HEE.]

This tree is well known in this country by the name of white walnut. The inner bark of this tree and especially the root affords one of the best purgative medicines with which I am acquainted. In bilious fevers, bilious cholics and in most cases where an active purge is required, some physicians object to this article on account of its drastic and stimulating properties, but it may be easily rendered one of the mildest and yet one of the most certain cathartics now in use by combining it with gulver root. The manner of extracting the virtue of this bark is by boiling it in water for several hours and then strain out the bark and boil the decoction to the consistency of molasses or pills, it desired, from three to five of these pills when fresh will operate on the bowels of most persons; if the gulver is to be added, put the root in and boil it with the butternut bark. White walnut forms an ingredient in Foreman's anti-bilious pills, and these pills are used by the Cherokee Indians in almost all cases where a purge is thought by them to be necessary. Butternut may be administered either in pills, extract, cordial or syrup. Its medical virtues, as I have before told you, are confined to the inner bark, and the proper time for getting it in the full possession of its virtues, is about the month of June--

the bark at this time in the year is considerably more powerful than at any other season.

RHEUBARB.--(Rheum Palmatum.)

Rheubarb is a native of some parts of Asia and of the East Indies, but is now cultivated both in Europe and America, for medicinal purposes. The root kept in the shops for sale, is imported from Russia, Turkey and the East Indies, but that which is cultivated in our own gardens, is equal if not superior to the best imported. The greatest inconvenience which attends the cultivation of this root is, the great length of time which it requires for it to come to perfection. Those acquainted with its cultivation say, that it ought not to be used until it is from six to ten years of age. Its cultivation is by no means difficult it is merely to sow the seeds in a light soil in the spring, to transplant the smaller roots the next spring into a light soil, well trenched; set the small roots about three or four feet apart. The third year the plants will produce the flowers; but the root is not fit for use before the fall of the sixth year, and will not have attained the full power of its virtues before the tenth year; it may, however, be used after the fall of the sixth year, with tolerable success as a purge. The proper time to take up the root, is in the fall after the leaves decay, or in the spring before they put forth. When taken up, the roots should be washed clean and the small fibers together, with the external rind pared off, after which, they must be carefully hung up in a dry place, no two touching each other, lest they mould, until they are completely dried, which will require from six to twelve months.

Rheubarb is one of the mildest, best and most pleasant purgatives now in use, with its purgative properties it is also astringent and strengthening; in this particular it differs from most cathartics. It is superior to nearly all purges for another reason, it may be taken with opium and act on the bowels as well as if taken without it. This is a vast advantage, where purging would be attended with great pain, it may be relieved by opium, and the rheubarb is left free to act on the bowels.

It may always be given with safety in all cases of extreme weakness, where a purge to open the bowels becomes necessary, and violent and severe purging would be highly improper. It is a valuable purge for children, even at a very early period of life and in every situation where their bowels become disordered, particularly in dysentery or lax. It is also a valuable purge for grown persons laboring under this complaint.

There are various modes of administering Rheubarb, such as in tincture, which means steeping it in any kind of spirits, in tea, &c. But the best and most certain method of giving this root is in fine powder. A dose for a grown person is from a tea to a table-spoonful. It may be used in tea or decoction, but by being heated it looses some of its strength or purgative properties. A valuable tincture or bitter, for persons afflicted with asthma, may be made as follows: Take of the root of Rheubarb pulverized, one ounce, of cinnamon bark one ounce, and one ounce of cloves, put all three of these articles into a quart of peach brandy, let it digest eight or ten days, shaking it well each day. This tincture may be used as other bitters, according to the strength and constitution of the patient.

PEACH-TREE.--(Amygdalus pesica.)

This tree, so common in all parts of our country, not only affords us a pleasant, delicious and wholesome fruit, but also furnishes us with some very valuable medicine.--Medical virtues of great value are to be found in the bark, leaves, blossoms, kernels, and gum. The flowers, if gathered in full bloom, and dried in the shade, are equal, if not superior, to either the imported or the American senna, in all cases in which it is useful, either among infants or adults. A tea of either the bark, leaves or flowers, will purge the bowels freely and without the least griping. Dose for an infant, is a tea-spoonful every half hour until it operates--for an adult, it must be taken in larger quantities; they also act as a purgative when taken in syrup. The syrup is prepared by boiling the tea or decoction of the bark, leaves, or blossoms, over a slow fire, with an equal quantity of honey, molasses or sugar to the consistency of syrup. The bark, leaves, and flowers, all or either, made into a strong tea, and taken a gill every hour until it operates on the stomach, bowels and skin, (for taken freely in this way, it will puke, purge and sweat the patient) has frequently thrown off bilious fever in its first stage, without the aid of other remedies. A decoction of the bark, leaves, or blossoms, sweetened with honey or sugar, is excellent, given to children a tea-spoonful every half hour until it operates for worms, hives, diseases of the skin, fevers, &c.--The gum of this tree will answer all the purposes for which gum arabic is used; it forms one of the best injections now in use for dysentary or flux.

The kernels taken from the peach stone is a very powerful tonic, and may be used alone or combined with other articles in cases of extreme debility. Children that are in the habit of eating peach kernels are seldom afflicted with worms to any extent. These kernels tinctured in brandy in proportion about four ounces to a quart, form a powerful tonic, and will be found beneficial in most cases of debility. It is very good for the Whites or Flour Albus in females:--the dose is a tea-spoonful two or three times a day.

BLACK ROOT, BRINTON ROOT, GULVER ROOT, BOWMAN'S ROOT.--(Leptandria Alba.)
[TE-NAH-TAH-NA-SKEE.]

This root is generally found growing in low wettish lands near streams and open glades or plains. It has a dark colored. perrenial root, which grows from a long woody candex or head. Several stems spring up from the same root--they are round, hairy and generally grow from two to four feet high--they are branched towards the top, the branches bearing on their tops a spike or tassle of white, crowded flowers. The leaves are long, narrow, pointed, and their edges indented with unequal teeth, growing in whorls of four or five at a joint. I am authorized from personal experience to give this root a high recommendation as an efficient purge, operating with mildness and certainty, without producing that depression of the living powers which so commonly result from the operation of purgative medicines. It forms an ingredient in Foreman's anti-bilious pills, and in this preparation renders mild the exciting properties of the butternut, and enables us to enjoy at once the active properties of the butternut, and the antiseptic properties of the brinton root.--This root appears to be peculiarly adapted to typhus and bilious fevers--it is with us the most appropriate purge to carry off the tarry, morbid matter from the intestines in these complaints. The dose is a heaping tea-spoonful in a gill of boiling water, repeated in three hours, it seldom fails to operate in that length of time. This root may be made into pills if preferred. It also forms our favorite emetic when combined with Indian physic, or American ipecacuanha, in most cases where an emetic is required.

This root is also a diaphoretic, tonic and antiseptic, which properties added to its cathartic powers, render it so valuable to evacuate the contents of the bowels in fevers. It is excellent in tincture for chronic or lingering complaints, where persons have recovered imperfectly from fevers, agues, dropsies.&c. This root may be used either green, dry, in tea, powders, pills, or tincture.

AMERICAN SENNA--(Cassia Marilandica.)
[AH-SUR-GEE-CAH-TEE-SAY-TSE-YOU-STA.]

This valuable plant is found in most parts of the United States, growing generally in rich soils, and near streams. It has a black, woody, crooked, fibrous perennial root. Several stems rising from one root to the height of from three to six feet; they are round upright and nearly smooth. Its leaves are alternate, large, and composed of many small leaves growing in pairs on one central stem or petiole.--The flowers are of a bright, yellow or orange color forming a square cluster at the top of the stem. The fruit consists of long pods, a little swelled at the seeds, and bears a slight resemblance to the locust pod, though not near so broad.

The American senna is cathartic, and is valuable among children or infants where a purge is necessary, it may be taken combined with sweet fennel seed to great advantage. Take a half ounce of the leaves and put them into three gills of hot water, take a table-spoonful for a dose every two hours until it operates. If the bowels be hard to evacuate, it may be taken in larger quantities. Many white physicians use the American in preferrence to the imported senna. For an adult it may be prepared and taken in the same manner as directed for children only in larger quantities.

MAYAPPLE, MANDRAKE.--(Podophyllum Peltatum.)
[OO-NEE-SQUA-TOO-KEY.]

The Mayapple is a well known plant growing in most parts of the United States, in shady moist lands. It has a perennial root, long, round and jointed, with many fibres or small roots issuing at each joint, the root runs horizontal in the ground, stem erect, smooth, round, from twelve to eighteen inches high, of a yellowish green cocolor, stem forked, each branch bearing a single large leaf at the top, between which in the fork when it is in bloom there is a single white flower, which is succeeded by a yellow acid fruit. The fruit of this plant is good for food. The leaves are poisonous, and its medical virtues are wholly confined to the root. The proper time for gathering the root is late in the fall when the leaves begin to die. It should be carefully dried in the shade, and used in the form of powders.

The American May Apple is an excellent, gentle, and effective purge, when properly used. It is acknowledged by many white physicians to be far superior to the jalap obtained in their shops; it operates more gently as a purge than jalap, and a much longer time. It is also preferable to jalap in other respects, it is less nauseous and more easily taken; less irritating to the stomach and bowels, and may be more easily used by delicate females and persons having weak and sensitive stomachs. It is an excellent article in intermittent fevers, it is also good in dropsy, pleurisy, and in incontinuence of urine. The dose is from half to a whole tea-spoonful of the powders, in very large doses; it operates with activity and power. If griping is apprehended, mix it with an equal quantity of gulver-root.--The index will refer you to a full description of this root. The Indians frequently roast the May-Apple root, and use it in this way; this renders it more mild and less drastic in its operations. A few drops of the expressed juice, put into the ear, is a valuable remedy for deafness. The powdered root is good to cleanse foul and ill-conditioned ulcers it destroys the proud flesh without injuring the sound; it also removes the morbid matter, and promotes the exfoliation of carious or rotten bones. The proper manner of applying it to ulcers and sores, is by sprinkling the powder once in from two to four days.

MULBERRY TREE--(Morus Nigra.)

The Mulberry is too well known to need description; it grows in great abundance in almost all parts of North America. The tree bears a very delicious fruit, which in common with many other fruits, possesses the property of quenching thirst, abating heat, and opening the bowels as a gentle laxative. A syrup made of the ripe fruit is one of the most innocent, agreeable and certain purges in our knowledge for infants of costive habits.

The inner bark of the root of the common black mulberry-tree, in doses of from half to a whole tea-spoonful of the powder operates as an excellent purgative.

A strong decoction in the bark made into soup, with an equal quantity of molasses taken in doses of a wine glass full, not only proves an excellent purgative, but it is also useful to expel worms, particularly the tape worm. The inner bark of the root digested in whiskey, makes a valuable laxative bitter.

FLUX WEED, WITCH WEED.

This is a small weed, growing from six inches to a foot high. The leaf is stiff and full of little stickers or spines all around the edges of it--it bears a small prickly burr which contains the seeds. The seeds are black, and are the part used for medicine. They are a very gentle laxative and anodyne. The mode of preparing them for use is to boil a gill of the seed in a quart of new milk, and of this decoction give a gill every half hour until it operates on the bowels or relief is obtained. This preparation is an excellent remedy for dysentary, diarrhoe and bowel complaints generally, giving immediate relief in almost all cases in which it is taken--in obstinate cases it will be well to add a few drops of Laudanum to the above decoction.

BUCK THORN--(Rhamnus Catharticus.)

Buck thorn is a shrub or bush found growing in the woods and about hedges--it generally grows to the height of from ten to fifteen feet--it flowers in June and the fruit ripens in September. The fruit when ripe has a faint disagreeable smell, and a sickening bitter taste. Both the bark and fruit of this shrub is a very powerful cathartic: it is very active and drastic in its operation, unless used in combination with other articles to moderate its effects--the berries is the part generally used for the cathartics, and when taken alone, they produce griping sickness and dryness of the mouth and throat,--leaving a thirst of long continuance. A decoction of the bark of this shrub, used as a wash is a certain cure for itch--it is good for sore or inflamed

BENNE PLANT.--[Sessaman Orientale.]

This plant is much cultivated in the gardens of the middle and southern states--it is a native product of Africa; but of late years, the seeds have been introduced into South Carolina and Georgia, by the African negroes, and is now cultivated to a considerable extent, and is highly prized for its medical properties.

It has a large, four cornered stalk, growing from two to four feet high, sending out a few short side branches.--Leaves are opposite, oblong, oval, and a little hairy.--Flowers are small, of a dirty, white color, succeeded by the seeds which ripen in the fall.

The leaves and seeds afford a valuable mucallagenous substance in decoction, or infusion; that of the seeds is oily. This infusion of the decoction is valuable in flux, dysentary, cholera-infantum or any other disease of the bowels. The seeds yield a greater proportion of oil than any other known vegatable. One hundred pounds of the seeds is said to yield ninety pounds of oil. This oil is a good mild cathartic medicine--and is much more pleasant to the taste than castor oil; it is said to keep many years without contracting any rancid smell or taste; it is also said to answer all purposes of salad oil. It is mild in its operations, and as to the dose, it should be taken until it produces the desired effect.

ALOE OR ALOES.--(Aloe perlivata.)

Aloes is distinguished into three kinds or species--as caballine, socottorine and hepatic. The two last are the best for use. Cabalina is called horse aloes. The socottorine aloes is the resinous product of a plant growing in the Indian Ocean. It has a dark yellowish red color, a glossy, clear surface, and is in some degree pellucid; it is easily pulverized, and when reduced to a powder, is of a bright golden color. Its taste is bitter and disagreeable though accompanied with an aromatic flavor.

The hepatic aloes is so called, because it is said to have a more direct and specific action on the liver, than the other kinds have. It has a strong disagreeable smell, and an intensely bitter sickening taste, accompanied with but very little, if any of the aromatic flavor of the socottorine.--Aloes is a valuable medicine, and may be used in pills, powders or tincture. It is said by white physicians to be one of the best correctors of the bile or biliary system.--It is also good for worms; from two to four grains will act on the bowels.

CLASS NO. III.

STIMULANTS.

Stimulants are of a class of medicines, which excite a new and stronger action in the system, or in some part of it, in order to, overcome an existing one, which is morbid, or two languid; or so excite the operation of an obstructed one. Those which produce permanent, and enduring effects are called *Tonics.* Stimulants are distinguished into two classes those which produce a universal excitement throughout the system, are termed *Diffusable* Stimulants, and those which are employed to excite in some particular organ or organs, are called *Local Stimulants.*

GARLIC.--(Allium Sativum.)

This valuable article is cultivated in gardens for its medical virtues and is two well known to require a description.

Garlic is a stimulant, carminative, diuretic, anti-scarbutic expectorant, and slightly cathartic. As a stimulant, it is both powerful and diffusable, and on this account is very useful for persons of cold phlegmatic habits. It forms an ingredient in Foreman's Phthisic mixture. The syrup increases the appetite, assists digestion, removes flatulence, promotes perspiration and has long been esteemed as useful for scurvey, dropsey and asthma. A very respectable white physician, of the old school, asserts that twenty cloves of Garlic, taken one of a morning, well pounded and mixed with common brown sugar, will cure any common case of asthma.

Garlic applied to the soles of the feet, says a respectable white physician, "exceeds any other application to produce a revulsion from the head," and on this account, the garlic poultice to the feet and the syrup taken internally, is a most excellent remedy for collection of humors in the brain. It is also very good for croup, sore throat, for this apply the poultice to the feet, and annoint the throat and breast with an ointment made by bruising the garlic and adding some lard. In malignant sore-throat the poultice applied to the throat also, will be found of service.

Cotton or wool, wet in the juice of garlic and put in the ear and renewed three or four times a day, is valuable for deafness, it has often removed it when other remedies had failed.

The Garlic poultice, applied to the feet in the low stage of accute disorders, and nervous fever, is good to raise the pulse, relieve the head and increase the general action of the system. Garlic given in tincture to children of a morning, will prevent worms, cholics, &c. The juice or syrup is far preferable to Bateman's drops, paragoric and many other articles of the kind so often given to infants for the cholic; in this case, a few drops of the expressed juice or syrup should be given every morning.

The proper manner of making the Garlic poultice, is to mix equal quantities of bruised Garlic and crumbs of bread, moistened wit vinegar. The Garlic, when applied alone, will draw a blister.

CALIMUS, SWEET-FLAG.--(Ocorus Cabimus.)

Calimus grows mostly in low marshy places, and in shallow water. It has long, sword-shaped leaves, resembling those of the flag, only they are much narrower and of a brighter green, they put out from the root without a stock, in the same manner as those of the Flag. Its root has a strong aromatic smell, and a warm bitter taste.--The flavor is improved by drying the root, it possesses stimulant and stomachic virtues, and is an excellent article in flatulent cholic, for both children and grown persons; for this purpose, grate or pound the root and put into water, or make into tea. It is recommended by some for White Swelling, prepared as follows.

At the commencement of this disease, make a strong decoction of the root of white willow, thicken it with flour and apply it to scatter. If this should fail and the swelling increase, when it becomes ripe lance it deep and let it run. Then take equal quantities of brimstone and calimus root, both finely pulverized and put them in a dry gourd prepared for the purpose. Then take of common lead that will make as much when melted as you have of either of the other articles, melt the lead and pour it slowly

into the gourd, at the same time shaking the gourd; the gourd must be well shook until the contents become cool. The above process will convert the lead and calimus into a powder resembling gunpowder. Take of this powder and introduce it as far in the orifice as convenient, and then apply some of the same externally to the affected part. The above treatment is said by some to be a never failing remedy for White Swelling. I have never tried it personally, as the mode of treatment laid down under the head of White Swelling in this book, has never failed to fill my most sanguine expectations. Calimus is also good to expell worms when taken repeatedly.

MOOR-WORT.--(Andromeda Marienna.)

Moor-Wort grows plentifully in many parts of the Southern States. A strong decoction of this plant is highly esteemed as a stimulating wash, and is very useful in ulcerations of the feet, such as toe-itch, ground-itch, a complaint very common and troublesome among the blacks in some parts of the Southern States. It is also good for ulcerations on other parts of the body, and forms an excellent wash for indolent ulcers, as it stimulates them and disposes them to heal.

BLUE FLAG, GLEET ROOT.
[OO-WA-SCO-YOU.]

Blue Flag Generally grows in low situations or flat lands and near strsams. It is called by the whites Blue Flag or Wild Flag, but by the Indians it is called Gleet Root, from its great efficacy in the cure of Gleet. Its leaves are of a deep green color, and in shape resemble those of the calimus, they grow to the height of from six to eight inches, it blooms in June or July; flowers are blue, with a bright yellow tinge in the centre--the root lies very shallow, being seldom entirely covered with earth--on the main root grow several small knots or lumps about the size of a small bean, they grow from one to five or six inches apart on the main root. The root is the part used for medicine, and is one of the most permanent stimulants with which I am acquainted, and is generally useful where articles of this kind are needed. A decoction made by boiling this root with an equal quantity of dew-berry brier root, is not surpassed by any article in the Indian Materia Medica in the cure of Gleet. It will also cure the Venereal.

This root pulverized and stewed in hog's lard, sheep suet and beeswax, forms a most excellent salve for ulcers, particularly cancerous ulcers; combined with other articles, it forms a valuable salve in all cases where a drawing salve is needed, but is too severe in most cases when used alone.

SAGE.--(Salvia.)

Sage is cultivated in most gardens for its medical virtues and for culinary purposes. An infusion or tea of the leaves sweetened with sugar or honey, is good in colds, coughs, nervous, debility, weakly females and persons of phlegmatic habits. A syrup or wax made by stewing sage leaves in honey until the strength is extracted, affords present relief in Asthma. I do not pretend to say that it will effect an entire cure for the disease, but it will relieve the spasm or fit immediately, and give ease until other remedies can be had. If the honey is not at hand, make a very strong tea of the leaves and put in it a teaspoonful of the flour of sulpher, this will also give temporary ease.--It is a mild diaphoretic and is valuable in all cases where such articles are required.

MISTLETOE--(Viscum.)
[OO-TAH-LEE]

Mistletoe, which is sometimes called Misseldine, is an evergreen which groes on several kinds of trees. That which grows on the oak is best for medicine. It is good for epilepsy or fits--for this purpose gather the mistletoe about the last of November or first of December, dry and pulverize it, and preserve it in well corked bottles.

SEVEN BARK,
[TE-TAH-NA-TAH-WA-SKEE.]

Is a shrub or herb growing mostly in low, wet, thin soil, its height is from eight to ten feet, it is covered with several coats of thin bark, which always appear to be scaling or pealing off, has large tough leaves, flowers white, appear in May and remain on the bush the principal part of the year. The bark and leaves have a very pungent, acrid taste, somewhat similar to that of prickly ash.

The inner bark and leaves are the parts used. It is stimulant, antiseptic, and antiemetic--valuable applied to ulcers, tumors, sprains, &c. Also taken in tea to stop vomiting.

CAMPHOR TREE--(Laurus Camphora.)

The article known in this country as gum camphor, is the product of the camphor tree, which is a native of Japan and grows in great abundance and to a considerable size in the forests of that country. The branches, trunk and roots, all contain the gum. It is separated from them by a process called sublimation, which is something similar to distillation. The tincture of camphor is a very common family medicine, and is certainly one of the best common-place medicines known to me. It is a valuable sweating medicine in all cases that require it, such as colds winter fevers, &c. It is also an excellent article in spasmodic affections. It is good for females in almost all hysterical or nervous diseases, and for nervous head-ache, both snuffed and applied externally. It is useful as a stimulant in sickness, fainting, &c., and as an anodyne in cholic, cramp &c. When persons are choaked with worms a dram of camphor will give immediate relief. It is excellent for strains and bruises, either in man or beast. After the stimulus ceases, which has been produced by the use of camphor, the person is apt to feel a disposition to sleep, without experiencing any bad effects from its use. It leaves no disagreeable effect in any way, if but a due portion be taker.

INDIAN TURNIP--(Arum Triphyllum.)
[TOO-YAH-STEE.]

This root has been in high repute among the American Indians, time immemorial. It grows in most States in the Union, in shady woods where the soil is light and rich. It has a perennial root, which in its shape, bears a great resemblance to the common turnip, though it is smaller; externally it is dark and wrinkled, internally it is white; its stalk grows to the height of from one to two feet and is of a reddish purple color, the leaves are three in number of a roundish or oval form, and one flower of the same color of the leaves, succeeded by a roundish cluster of berries of a bright and beautiful scarlet color when ripe.

In its fresh or green state, the Indian turnip has a powerfully acrid, biting taste; it is stimulant, expectorant, diaphoretic, and carminative. When dried it loses much of its intolerable pungency, together with much of its virtue. Some acrimony or sharpness should be perceptible to the throat and tongue, or the root has lost its powers, and should not be relied on. The root pounded to powder and mixed with honey, a tea-spoonful for a dose, two or three times a day, is good for colds, dry coughs, phthisics, &c.; it is also good in whooping-cough, consumptions, &c. It may be taken by boiling the fresh root in sweet milk until tolerably strong, and take a tea-cupful morning and night, the dose must be varied to suit the strength of the patient.--An ointment made by stewing the green root in hog's lard is valuable for scald head, keeping the head cleansed with soapsuds once a day. This ointment is also good for ringworm, tetter-worm, scrofulous sores, &c. Indian turnip is one of the most valuable expectorants with which I am acquainted, and may be used alone or combined with other articles in forming expectorant compounds.

GINGER, RACE--WHITE, BLACK GINGER.
(Amomum Zingiber.)

The article known in this country by this name, is the root of a perennial shrub, which is extensively cultivated in both East and West Indies. Race is a term applied to the root to distinguish it from that which is ground or pulverized. The white ginger is that which was washed and scraped before drying and the black is that which was washed only without having the external, dark bark taken off the root.

Ginger is a warm, stimulating aromatic, and is a most valuable article in the practice of medicine. It is good for colds and cholics in tea--for cold, weak constitutions it is a good tonic. It is good for females at their monthly periods, if the discharges be too scant, and for lying-in-women. A weak tea of ginger is good for infants when they are inclined to be hivy or colicky. It is also good in all cases of looseness and weakness of the bowels or intestines; it does not heat the system as much as the different kinds of pepper, but is much more durable in its effects. Externally it is a very valuable ingredient in stimulating poultices. It is one of the best articles to relieve heart-burn in pregnant women with which I am acquainted--for this purpose they should chew the root, or pulverize it and take it until relief is obtained, there being no danger whatever in its use.

In purchasing ginger for medical purposes the root is to be preferred, as that which is brought on in the pulverized state is often prepared of unsound or worm-eaten roots, or adulterated with other articles.

BLACK PEPPER--(Piper Nigrum.)

The article known in this country by the name of black pepper, is the fruit of a tree which grows spontaneously in the East Indies. The berries are gathered and dried before they are ripe, and to this circumstance, they are indebted for their black color.

Black pepper is much used as a condiment in cookery.--It is stimulus and slightly astringent, and may be employed as a substitute for cayenne, or red pepper, where they cannot be had.

RED PEPPER--CAYENNE PEPPER.
(Capsicum Annuum.)

Cayenne is a native of the tropical climates, but it is now cultivated in temperate ones also. Of this article there are several species, all possessing the same medical properties to a greater or less extent. That growing in Africa is considered the best or strongest, and is generally called African Cayenne.

Cayenne is a very powerful stimulant, and is valuable in colds, cholics, &c. It forms an ingredient in the Phthisic mixture. The red pepper poultice applied to the soles of the feet, in nervous or low fevers, is a valuable remedy, as it raises the pulse, and produces a revulsion from the head; it has good effects on poultices on gangrenous parts. The essence made by putting three or four pods in a half pint of whiskey and burning it one third away, is an excellent application to remove pains in the side or elsewhere, it should be applied externally to the pained part and bathed in well, or a piece of flannel wet in this essence

and kept to the pained part as long and often as it can be borne. It is also a valuable stimulant in animal poisons, such as snake-bites, &c. When the other remedies prescribed for snake-bite are not at hand, give red pepper in whiskey freely, until the pulse is raised, and repeat as often as the pulse sinks or becomes weak.

SAFFRON--(Crocus Sativus.)

Saffron is cultivated in gardens both in Europe and America for its medical qualities. It has a pleasant smell, and an aromatic, bitter taste; and when chewed it imparts a deep yellow to the spittle. It is a valuable article among children--good for hives, jaundice, red gum, and eruptive diseases in general.

BUTTON SNAKE-ROOT--(Liatris Spicata.)
[TOO-NOO-NOO.]

This root is a native of all the Southern States from seaboard to the Mississippi. It has a rough, perennial, fibrous root, and on the fibres grow little button-like knobs. Its stem is round and sometimes branched, bearing on the top a spike or tassel of scaly, purple flowers, which slightly resemble the shape of an acorn. This root possesses many medical properties. It is a warming stimulant, a dimetic, sudorific expectorant, carminative and anodyne. A decoction or tincture of the root is a valuable remedy in most cases of colic; it is also good for back-ache, pains in the limbs, dropsy, &c. It has a sharp aromatic and very bitter taste, and when chewed it produces a considerable flow of saliva or spittle. By many physicians of reputation, it is held in higher estimation than the seneka snake-root, which it very much resembles in its effects.

DWARF-BAY, MEZEREON--(Daphne Mezereum)

Dwarf-Bay mostly grows in shady woody places where the soil is rich, and may be found in great abundance near the Ohio river. The leaves are spear shaped, flowers put forth in the months of February and March, and are of a beautiful red or rose color.

The bark of the root is the part used for medicine. It has an extremely acrid, burning taste, and is so irritating that it cannot be used constantly, but may be used in small portions, and at intervals, regulating the time and quantity by its effects. This article is highly stimulant and diaphoretic, and is one of the most valuable articles in the Cherokee materia medica in the last stages of the venereal where the constitution has been impaired by the improper use of mercury in this disease, which is too often the case under the old system of practice. This article is then found most efficient in relieving nocturnal pains and removing what is called venereal nodes, This root may be taken in decoction either alone or combined with other articles, as may best suit the views of the patient.

BUTTERFLY WEED, PLEURISY ROOT, FLUXROOT, &c.--(Asclepias Tuberosa. (GOO-GOO.)

This beautiful plant is a native of every State in the Union, but is most abundant in the south and south western States. It flourishes best and grows to the greatest perfection in light, sandy, or gravelly soil, and is frequently found along fences and near stumps in grain fields. It has a large, crooked, branched, perennial root, of a light brown color on the outside, and white within, several stems rise from the same root, sometimes they are twenty or thirty in number, about the size of a pipe stem and stand in almost every direction, they are round, wooly or hairy and branched, rising from one to two feet high. The leaves are placed very irregular on the stalk, and are covered with a fine down on the lower side, thick or fleshy, and of an oblong shape. Its flowers appear in July or August they grow in terminal, corymbose umbels, and are of a most beautiful, brilliant, orange color, and is easily distinguished from all the flowers that adorn the fields.--This plant is often mistaken for the common silk weed.--There is, however, this difference between them, by which they may be easily distinguished; the flowers of the pleurisy root are of beautiful, bright orange color, while those of the silk weed are of a pale purple hue.

Few articles in the Indian Materia Medica maintain a higher standing for its medical virtues than pleurisy root The powdered root acts as a mild purgative on the bowels but it is more particularly and inestimably valuable in producing expectoration, or throwing off mucus from the throat and lungs, and in causing perspiration or sweating, when other remedies fail. This root possesses one remarkable power; given in proper quantities, it affects the skin, and produces perspiration or sweating, without heating the body, or increasing the circulation. It is a valuable article in diseases of the lungs generally. It is a powerful remedy in pleurisy, as may be seen under that head. It is a valuable article in the treatment of bowel complaints among children. Its use in a strong decoction, often gives relief to pain in the breast, stomach and intestines, by promoting perspiration, and assisting digestion. In feverish affections, proceeding from inflammation of the lungs, in colds recently taken, and in diseases of the chest generally, it is an excellent remedy. It may be administered, either in decoction or powder, of the pulverized root a tea-spoon-ful or more may be taken for a dose, and repeated as often as necessary; we generally combine it with the silk weed root, equal quantities; and if you wish to produce copious perspiration, without raising the internal heat, no article spoken of in this work will be better adapted to this purpose than the above compound.

WHITE SNAKE ROOT--POOL ROOT.

Pool root is found in great abundance in the western states, principally confined to dry upland soils, and to lands timbered with oak and hickory.

The root is small and fibrous, growing from six to four inches long, and of a dirty white color. Its stem rises from one to three feet high, angular and furrowed. The leaves are opposite, alternately supported on long foot stalks broad at the base, acute at the point with edges obtusely tentate or toothed. The flowers are white, and grow out in beautiful clusters.

The root is the part used for medicine, is stimulant, tonic and diuretic, and has a warm, aromatic taste. This root may be used either in decoction or tincture, and is valuable in fever and ague, and will seldom fail in effecting a speedy cure, if the stomach and bowels have been properly cleansed, previous to its use. It is also given for gravel and diseases of the urinary organs generally. It may be used to advantage in most cases, where a stimulant is required.

ALLSPICE TREE, PIMENTO TREE.
(Myrtus Pimenta.)

The Pimento Tree is the spontaneous product of Jamaica, one of the West India Islands. What we call Allspice, is the fruit of the Pimento tree it is plucked from the tree before it is ripe and dried in the sun. The proper name of this fruit is Pimento or Jamaica pepper; but its scent resembling that of a mixture of cinnamon, nutmeg and cloves, it has received the name of *All spice.* It is a warm aromatic stimulant, and is useful where gentle, stimulant stomachs are needed. It is a valuable astringent, stimulant for lying-in-women, whose discharges (called by physicans *Lochia,*) are profuse, but should not be used by those whose dischares are scant. It is also good for females at their monthly periods, when their discharges are profuse and weakning, but should not be used by such as are laboring under suppression or obstruction of the menstrual discharges.

CINNAMON TREE--(Laurus Cinamomum.)

This tree or bush is a native of the Isle of Ceylon, in the East Indies, but it is now cultivated in many of the West Indies. It grows to the height of ten or twelve feet, and is very bushy. Its leaves resemble those of the laurel, and when chewed, have the hot taste and aromatic smell of cloves. The article known in this country as Cinnamon bark, is the inner bark of this tree.

The bark is a useful and pleasant aromatic--it has a very pleasant taste, and strengthening to the stomach.--It is stimulant, stomachic, tonic and carminative, and is a valuable article for females in the latter stages of pregnancy, it may be used in decoction, powder, or combined with spikenard, digested in spirits and taken as bitters, its constant use for the last three months before delivery, will greatly strengthen both the mother and her offspring. It is also good to stop flooding either before or after delivery.

MOTHER-WORT.--(Leonurus Cardiaca.)

Mother-Wort mostly grows in water places; it flowers in July and August. The flowers are white on the out side, and purplish within, growing in thorny whorls. The leaves are opposite; two to each whorl; and have a strong disagreeable smell, and bitter taste. An infusion of this plant is a stimulant, reviving, cordial bitter. It is valuable in nervous and hysterical affections, and when taken at bed time, it procures a quiet, refreshing sleep, even where opium and laudanum have faild. It has also been used to great advantage in faintings, and diseases of the stomach. The quantity taken must be regulated, according to the strength of the patient, and the effect wished to be produced.

WILD GINGER, HEART SNAKE ROOT. (Asarum Candensis.) [TSE-YOU-CAH-LE.]

This herb is found in most states, but is most abundant in the South. It generally grows in rich, shady, moist, wood lands. It has a round, fleshy, jointed, perennial root, which runs horizontally in the ground having many fibers; its colors is a bright outside. Its leaves are radical, round, hazy, veined; two from each root; supported on long foot-stalks, so close the ground as to be scarcely perceivable.

The root of this plant is a warm, powerful and diffusible stimulant, and on this account, it is valuable for colds, coughs and female obstructions. It may be used in decoction, tincture or syrup.

This root combined with star root and a small quantity of sampson snake-soot, or a small quantity of puccoon digested in common spirits, forms an excellent bitter for females, whose menstrual discharges are scant, and painful, or entirely obstructed. The root is also a valuable Anthelmintic, which means the property of discharging or expelling worms. When taken as a medicine, it should be boiled in sweet milk, and drank freely. It is valuable in warm fevers, and in most cases, where a diffusive stimulent is needed.

A snuff made of the dried leaves, finely pulverized, is very good, when snuffed up the nose for the head and eyes.

PINE.--(Pinus.)
[NOH-TRE.]

The common pine of our forest, may well be ranked among the most beautiful and useful trees with which our

country is blessed. Beside the many uses made of it by mechanics for lumber, &c., its buds, bark, rosin and roots, possess medical virtues that are almost innumerable.

It is stimulant and acts gently on the bowels as a laxative. The buds or inside bark steeped in water is excellent for hard, dry coughs, two or three glassfuls a day should be taken warm. A daily use of the inside bark taken in decoction is valuable in old bowel complaints. The spirits or oil made by distilling the turpentine or rosin which exudes from the tree where an incision is made; when taken internally, is one of the most active and diffusible stimulants within the compass of medicine. In cases where the bowels are obstinately constipated or bound, it is the most certain remedy with which I am acquainted. It is also useful in worms, hysterics, rheumatism, cholics, gout, weak back or kidneys, and in the treatment of child-bed fevers.

The rosin in its natural state or as it exudes from the tree, is good in spirits for old rheumatism, and weak back and loins, or taken in pills it will answer a similar purpose. It forms an ingredient in the pill for dropsey. Many other uses are made of the pine rosin in preparing salves, ointments, &c. The rosin obtained by boiling the root forms one of the best strengthening plasters in the world. The index will refer you to the proper head for making and applying this plaster.

ROSIN WEED.

This singular, and valuable plant, is found growing principally in the north and north-west, in those sections of country that are destitute of pine, the place of which this plant appears expressly intended to supply. It grows in rich ground, and is a very large rich looking plant.--The roots are large and grow deep in the ground the stocks rises from three to six feet high, rough, about the thickness of a man's thumb and crooked towards the top, the leaves are large, partly radical, the remainder grow irregularly up the stalk, and are of a dirty or ash colored green. Whenever the stalks is broken or the bark taken off, a rosin exudes from it like the opium does from the poppy, and adheres to the stalk in dry brittle lumps.--So nearly does this rosin resemble the pine rosin in color, taste and smell, that a person, not intimately acquainted with both, cannot discriminate between them.

The rosin may be used in all cases in which the pine rosin is recommended, for its effects on the system bear so strong a similitude to those produced by the pine rosin that they appear the same. The roots digested in spirits is good for females troubled with the whites (flour albus.)--The pine rosin is also good for weakly females. The pine and rosin weed is never found growing in the same section of country, I believe; and they seem expressly intended to supply the place of each other, and furnish both the cold and the warm climates with a medicine at once sate and valuable in the treatment of many of the diseases to which the inhabitants of both countries are subject.

WILD-WET-FIRE, BLISTER-ROOT, CHICKEN PEPPER.

This plant is found growing on the banks of creeks, and spring branches near the edge of the water. Its root is white, many fibers issue from the caudex or main root, generally two but sometimes three leaves put up from a root, each leaf is supported by a foot-stalk from one to three inches high, the leaves and foot-stalks are covered with a kind of hair or furze of a green color, having whitish spots or pides, they are indented around the edges, and sometimes grow to the size of a silver dollar, but are generally much smaller. The root is the part used bruised and applied to the skin it will draw a blister much quicker than Spanish-flies. When a blister as large as the hand is desired, apply a plaster of the bruised root as large as a dollar; when the blister is drawn, annoint the edge with oil or lard to prevent it from spreading, and dress it with cabbage leaves, brier leaves, or plantain leaves, prepared as is common for a blister drawn with Spanish-flies.

WHITE POPPY.--(Papaver Album.)
AND MODE OF PREPARING OPIUM FROM IT.

The cultivation of the Poppy, the Drug Opium and its medical properties were equally unknown to the *Aborignnees* of our country previous to their acquaintauce with the whites. They frequently use Opium as a medicine since it was taught them by the

Pale Face, and many of their physicians believe it to be a most excellent remedy in many cases. Opium is obtained from the White Poppy.which is said to be a native of Asia, but is extensively cultivated in many parts of Europe, it has been cultivated in the United States sufficiently to prove that the soil and climate are as well adapted to its culture as any part of the Globe. The leaves, stalks, and capsules of the Poppy abound with a milky substance, this substance, or juice, is obtained by making incisions or cuts lengthwise on the capsules or pods about sunset, during the night the milky juice will exude from the pods through the cuts, and adhere to the sides of the incisions; on the following day it must be collected into an earthen plate. This is best done by a thin iron scraper made for the purpose. When thus collected, you are to work it in the sun with a wooden paddle until the juice becomes sufficiently thick, then make it into cakes with the hands and wrap it in the leaves of the Poppy, and put it into glass jars or bottles; if these are not at hand, wrap it in a hog's or beef's bladder, and it will keep as long as desired. The operation of cutting the Poppy pods as before directed, may be repeated every evening as long as the pods will furnish the milky juice. The best time to commence making the incisions, is when the pods are a little more than half ripe, the cuts are most easily made with a small knife having three blades, the middle blade being the shortest. The cuts should be shallow, not extending quite through the hull. This is the whole process of making Opium. There are two kinds of Opium. There are two kinds of Opium imported into America for sate, distinguished by the name of the *Turkish* and *East Indian* Opium. The Turkish Opium is the best it is more solid and compact, and when broken presents a smooth shining fracture. When good it is of a reddish brown color. When taken into the mouth, it produces a nauseous bitter taste, leaving a strong biting impression on the tongue and lips. It is very heavy and easy pulverized. It has a strong narcotic smell, and the cakes are wrapped in poppy leaves, frequently having on them many red-colored pieces of the capsules or hulls, which are indicative of its good quality. The East Indian Opium, which is not so good as the Turkish, has not that peculiar narcotic smell. It is much blacker, more nauseous and less bitter. It is not near so hard but more waxy and tenacious, and when broken, has not that uniformity of appearance which is found in the Turkish, but frequently has other particles interspersed through it. Laudanum is made by dissolving an ounce of Opium in a pint of good spirits of any kind; it is generally fit for use in six or eight days. Twenty-five drops of laudanum are equal to one grain of Opium. Opium and laudanum given in small doses act as stimulants; in larger doses they produce sleep and relieve pain; in an over dose, when the person is not in the habit of using it, the consequence will always be fatal, in this ease it produces vertigo, stupor, tremors, convulsions, insensibility, succeeded by a total deprivation of muscular strength, when death usually closes the scene.--When it is used as a luxury, (which is the case in China and some other parts,) and its use persisted in, it effects the physical system in a manner similar to the excessive use of intoxicating liquors. There is however this difference between the effects produced by spirituous liquors, and those produced by Opium. The excitement produced by spirits, are more acute and powerful while it lasts, but are of shorter duration than the effects produced by Opium. Spirits disorder the mind, unsettle and cloud the judgement and deprive us of our intellectual self-posession; while Opium on the contrary, soothes and tranquilizes the system, arouses all our dormant faculties, and produces just equipois between our intellectual strength and sensibilities.

Opium is a most powerful anti-spasmodic, and may be advantageously used to remove cramps, spasms, &c.-- When given to children, (if given at all) a half drop of laudanum will be sufficient for a child of a few weeks old. In highly inflammatory cases, opium should be seldom used, as it will tend to aggravate the symptoms. A common dose of opium for a grown person, is one gram, but the dose must be varied according to the age and constitution the nature and stage of the disease, &c. spasms or cramps, it may be employed in much larger portions with the happiest results.

LETTUCE.

The extract of the common garden Lettuce produces nearly the same effects as Opium. It possesses the power of allaying pain and producing sleep. As a stimulant it does not act so powerful as the opium produced from the Poppy, but has a tendency to repress the inordinate heat of the system, and to diminish the too frequent action of the heart, without producing those unpleasant effects which sometimes follow the use of Opium, by persons whose constitutions cannot bear the stimulus produced by the latter. The extract of Lettuce is obtained as follows: Take the leaves and stalks of either kind of garden Lettuce, when the plants are nearly ready to flower, bruise them well in a mortar and put them in a bag made of flax or hemp, then press them until they yield their milky juice--this juice is to be evaporated in flat vessels in the sun, or by placing them in boiling water, until reduced to the consistence of thick molasses, when it is to be bottled for use.

BLUE ROOT.

This very valuable herb is an evergreen, and is mostly found growing on the banks of small streams. The main root is about the thickness of a pipe-stem, of a purple red color externally, but when broken; the inside has a bluish cast; many small fibres issue from the main root, which are white. The leaves are supported by footstalks from two to four inches long, which put up from the root similar to the puccoon leaves; they are round, saw-edged, of a dark green color, and from three to five in number; they vary from the size of a dollar to twice that size. In the spring the flower stalk puts forth and grows from eight to ten inches high, bearing yellow flowers, which very much resemble the bloom of the common turnip. The root is the part used and is valuable in rheumatism and several other diseases as is fully shown in their different treatments.

CLASS NO. II.

TONICS.

Tonics are a class of medicines that increase the tone of the muscular fibres, and thereby strengthen the whole body. It has pleased the great Author of our being to furnish the vegetable kingdom with a a great many valuable articles of this class. A class of medicines designed to increase the tone and strength of the organic system, and invigorate the living power by improving the appetite, giving vigor to the muscles and to the digestive powers, must be of extensive utility.

Tonics may be distinguished into two kinds, viz: *Bitter Tonics,* and *Astringent Tonics.* The former are used to increase the tone and strength of the system generally, while the latter are peculiarly appropriate in the treatment of dysentery, diarrhoea, &c.

TAG ALDER.--(Alnus Serrulata.)
[TSA-TAH-NAH.]

Tag Alder is a perennial shrub, found mostly in low wet soils and along streams. It grows from eight to twelve and sometimes fifteen feet high, several in a cluster, the leaves are large, rather obtuse, of a dark green color.--This bush or shrub bears tags a little similar to those of witch hazle, from which it derives the name of Tag Alder.

It is an excellent tonic and is a safe, valuable and cheap article of family medicine. The bark, leaves, or tags may be used either in a decoction or digested in common spirits. It is excellent for women troubled with bearing down pains, either before or after child-birth.

Cloths kept wet with the decoction, and applied warm to painful swellings, afford much relief and generally scatters them. A poultice made of the inside, bark, leaves or tags, is an excellent remedy for strains and swellings, applied to swelled and pained testacles, whether produced by mumps or other diseases, it seldom fails to reduce the swelling and relieve the pain. The decoction drank freely is valuable in promoting the discharge of urine. The decoction or tincture is good for eruptions or diseases of the skin and particularly biles.

DOG-FENNEL, MAY WEED--WILD CAMOMILE.
(Anthemis Cotula.)

Dog-Fennel is a well known plant, abounding in every part of the United States with which I am acquainted, and is said to be a species of Camomile. It has a very offensive smell and a bitter taste.

This plant, although generally looked upon as one of the most offensive and useless weeds with which our farms are beset, is nevertheless a valuable medicine. It is an emetic, tonic, sudorific and anodyne, and may be advantageously used in colds, hysterics, epilepsy, dropsy, asthma, rheumatism and intermittent fevers, or fever and ague. A tea of this plant taken in small doses, as warm as it can be drank, promotes copious perspiraiton or sweating, and has often of itself, relieved persons afflicted with chills and fevers when other remedies had failed--and when properly taken and the bowels kept open by suitable medicines, it seldom, if ever, fails to effect a cure, in this disease. A poultice made by thickening wheat bran in a decoction of dog-fennel, is a valuable application in the treatment of inflammatory rheumatism. The bruized herb applied externally will draw a blister in a short time, equal in every respect to those drawn by Spanish flies.

BONESET, CROSS-WORT, INDIAN-SAGE, THOROUGH-WORT, THOROUGH-STEM. (Eupatorium Perfoliatum..) [TE-SCO-YOU-TE-NER-WO.]

This herb is found growing in marshes, swamps and wet meadow lands, throughout the United States. It has a crooked, fibrous, perennial root, running horizontal in the ground--several stems usually rise from the same root from two to four feet high, hairy, of a pale or grayish green color, branched to wards the top. The leaves grow opposite and are so formed as to have the appearance of being penetrated by the stem through the centre where they are broadest, and gradually tapering to a point.--The whole herb has a rough, wooly or hairy appearance. The flowers grow in dense corymbs or clusters on the top of the stems, and are of a dirty white color, and appear in the month of July.

This plant is possessed of powerful medical virtues. It is tonic, sudorific, stimulant, emetic, cathartic, antiseptic and diuretic. The warm infusion, given in doses sufficiently large to create nausea, produces the most copious and pleasant sweats, without increasing the fever in the least. The common mode of using the article is about a handful of the leaves to a quart of boiling water, when the strength is extracted, take about a half gill or wine glass full every hour or two. If the above dose be increased to twice or three times the quantity, and taken a little above blood heat, it will act as an emetic, emptying the stomach not only of its ordinary contents, but of the bile also. In intermittent and remittent fevers, it is far more beneficial and safe than the peruvian bark, for if the peruvian bark be administered when there is fever, its effects are dangerous, but the boneset may be given when there is considerable fever, with the most salutary effect, as its active sweating powers always tend to diminish the fever. The infusion given cold, a half gill taken every half hour, will produce purging and is an excellent article in obstinate constipation of the bowels. The warm tea is good in coughs, asthma and hysterical complaints. It is a valuable medicine in yellow fever.

This medicine acts well on the biliary or bile system, and also on the liver, giving them a healthy action, by which they are enabled to throw off all superfluous matter. It is the extract of this herb that constitutes Dr. Foreman's antiseptic pill, which operates so powerfully in assisting digestion. The leaves or flowers in powders, in doses of ten, fifteen or even twenty grains acts well as a purge.--This herb is also a valuable medicine in diseases of the skin. When boneset is taken as a sweat in cases of fever the stomach and bowels should be previously evacuated.--It sometimes increases the urine greatly, and has been used to advantage in cases of dropsy. In fever and ague it is an almost infallible remedy when the stomach and bowels are properly cleansed previous to its use.

SPIKENARD.--(Aralia Racemosa--Nardus Indica.)
[YOH-NAH-TSU-NE-YAH-STEE.]

Spikenard, usually called spignard, generally grows in rich, rocky grounds, both on mountains, in hollows, and in bottoms--it has a perennial root of a brownish yellow color, which divides from one caudex or head into several branches or distinct roots which are generally very long, and not quite so thick as a common finger--they are very tough, and when cut, broken, or boiled they yield a gummy substance of an aromatic smell and taste. This is the best part of the plant for medical purposes. Sometimes but one stem, and sometimes more rises from the same root

from two to three feet high; they are generally thick and of a purplish color, branched towards the top--leaves are biternate, (which means having three,) consisting of nine foliobs or small leaves. The flowers are of a yellowish white, growing in umbels, and produce small berries which bear some resemblance to the elder-berry. Both roots and berries may be used for medicine, and may be employed either in tea, syrup or tincture.

To do justice to this root, would occupy more space than the size of this book will allow for one root. It is tonic, diaphoretic, antiseptic, astringent and expectorant. This article is useful in coughs, asthma, and diseases of the lungs generally. By many physicians of high standing it is ranked amongst the master remedies for consumption. The manner of using it in this disease is as follows: Boil the roots until the strength is extracted, then take them out and strain the decoction, put it on a slow fire, and reduce it almost to thin syrup, sweeten it with honey an let it stand until it ferments; of this beer take a tea-cupful morning, noon and night. It is one of the best articles in the Indian Materia Medica for female weakness. For weakly females that are liable to floodings or slight menstrual discharges during pregnancy, it is the best article with which I am acquainted; for this purpose it may be taken in tea or bitters as the patient may prefer. For females afflicted with a weak back it is also good; for profuse menstruation, caused by a relaxed state of the system it is an excellent article. For females whose menstrual discharges are scant or irregular it is useful combined with rattle-root or wild-ginger. It is one of the most powerful strengtheners of the womb with which I am acquainted.--It is a fine application in fresh wounds, cuts. &c. The syrup in such cases should be made as thick as honey, and applied to the wound by means of dipping lint into the syrup and applying it to the wound. When prepared in this way it may be kept ready for use a great while, by adding rum or other good spirits, and is equal to the syrup made of white hickory-bark.

BLACK SNAKE-ROOT, VIRGINIA SNAKE-ROOT. (Aritolochia Serpentaria.)
[OO-NAH-STA-RES-TEE.]

The black or Virginia snake-root grows in great abundance in the different mountains in the United States, it is also found growing in rich river bottoms and on rich hill sides, generally in shady places.

This root has a small, bushy root, consisting of a number of small fibers matted together, issuing from one common head; it is of a brown color on the outside, and yellowish within, turning darker on drying. It has a slender, crooked stem, growing from six to ten inches high, bearing from three to seven leaves, which are long and heart-shaped at the base.

The root has a strong disagreeable smell, somewhat aromatic, and a very pungent and lasting bitter taste. Taken in strong decoction, is an excellent stimulus or tonic, and is beneficial in typhus fevers, ague and fever, &c., after preparing the stomach for it. In decoction, it is best; a handful of the roots to a quart of boiling water, taken in half gill doses, every half hour, until sweating is produced. It is also diuretic, and antiseptic, and has been used with advantage as a gargle in putrid sore throat. Taken internally, will stop mortification, and prevent putrefaction in the bowels. It may be used alone in tincture or compounded with other articles for bitters, and is valuable for persons of weak phlegmatic habits.

SAMPSON SNAKE-ROOT.
[OO-HAR-STEE.]

This plant grows mostly on dry poor grounds, in the woods. It has a perennial root; several roots issue from the main head about the size of goose quills, or hardly so large; and are supposed to bear some resemblance to worms. Its stems rise from eight to ten inches high, round, jointed; its leaves are opposite, smooth, oblong, few and of a beautiful, dark, green color. The flowers grow on the top of the stem or branches, are tatular, or cylindrical, of a dirty white, reddish, brown, or pale blue color, and never appear to be fully opened or expanded.

The root is the part used for medical purposes, and has a very pleasant taste. It is a very powerful and valuable bitter tonic, whether used alone, or combined with other tonic articles. It is an important article in the Indian practice, in all preparations for obstructed menstruation; it is an excellent article, in cases of cholic, indigestion, &c. It is diaphoretic, or sweating, and may be advantageously combined with other articles for this purpose, but it should not be used by females, in a state of pregnancy, or such as are troubled with profuse menstruation, as it will have a tendency to produce unfavorable symptoms. It may be used in decoction or bitters.

WILD CUCUMBER.--(Cicumis Agrotis.)

This tree is a native of the American forest, and is frequently known by the name of magnolia. In fertile soils, it sometimes attains the height of eighty or ninety feet.--It leaves are very large and beautiful, oval or tongue shaped; it produces a large fruit which bears some resemblance to the cucumber; it is from three to six inches long, about an inch or an inch and a quarter in diameter, and nearly all the way of a size. The fruit contains large seeds. About the end of the summer or beginning of autumn, it ripens or bursts open, and the seeds appear, being when ripe, of a beautiful red color.

The fruit, inside bark, and bark of the root all possess similar medical properties. It has a bitter aromatic taste, and when tinctured in spirits, makes a valuable bitter to increase the tone of the stomach. A free use of the tincture made pretty strong, is a good article for chronic rheumatism, particularly for persons of phlegmatic or weakly habits. Combined with dog-wood bark, it forms an excellent tonic for chills and fevers, or ague and fever. The tincture or decoction is an excellent remedy for obstructed menses, particularly where the general health is impaired by the obstruction; in this case it should be regularly taken, two or three times a day, until relief is obtained. In addition to its tonic properties, it acts gently on the bowels, when taken in sufficient quantities. The bark or fruit may be used fresh, or when first taken from the tree, but when properly dried, it is equally good. The inside bark or bark of the root and fruit, if dried, should be dried in the shade, allowing them fresh air, but excluding all dampness, such as rain, dew, night air, &c.

TANSY.--(Tanaceum Vulgare.)

Common tansy is a perennial plant, which is cultivated in most gardens in this country for medical purposes. It flowers in June and July.

Tansy is a warm bitter accompanied with a strong flavor, which is not very disagreeable. It is a valuable tonic, increasing the general strength of the organic system.--It is valuable in preventing abortions and miscarriages in pregnent women, either taken in spirits, decoction or infusion. It should be bruised and worn around the waist, and in the shoes next the soles of the feet in females who are predisposed to miscarriages, particularly about the time miscarriage generally takes place, or when symptoms of misfortune approach. The expressed juice alone or combined with rue, worm-wood, or either and sweetened with honey or sugar, and given to children of a morning is an excellent worm medicine. It is a good article in all cases in which a poultice of bitter herb is recommended in this book.

PRICKLY ASH, PRICKLY ELDER.--*(Aralia Spinosa.)*

Prickly ash is a perennial shrub or bush, growing mostly in rich bottom lands, to the height of ten or fifteen feet. The stem and branches are defended by sharp prickly spines. The bark is of an ash color, generally spotted or pided.

The bark and berries have a warm, pungent taste. It is tonic, diaphoretic and carminative. The bark of the root is the most active part of the shrub for medical purposes, digested in spirits it forms one of the best bitters for chronic rheumatism, and old venerial diseases now known, it also good for flatulent cholic. When given in inflamatory rheumatism, it should be given in decoction, instead of tincture or bitters; boil, say an ounce of the bark in a quart of water, of this drink a pint a day, divided into three equal portions, taken morning, noon, and night, it may be diluted or weakened with water to render it less pungent and more palatable.

A tincture of the berries, or bark is good to prevent the aching of decayed teeth, and for persons of weakly phlegmatic or inactive habits it is much better than the decoction.

SOURWOOD.

Sourwood is a native of the United States and is too well known to require a description.

It is a valuable tonic in dyspepsy or indigestion. The bark or leaves should be extracted, then strain the decoction and boil it to the consistency of molasses, and sweeten it with sugar or honey, the addition of a small quantity of soot will aid it in tranquilizing the stomach--it should be taken in teaspoonful doses, morning and evening. The sourwood molasses combined with sugar and British or sweet oil, is also a valuable remedy for phthisic or asthma, and for diseases of the lungs generally.

BLACK ASH-TREE.
[TSU-COH-NO-HUH.]

This tree is a native of the United States, and is to be found in great abundance in almost every part with which I am acquainted. The inner bark is a valuable tonic, particularly in chronic complaints of the liver. It acts as a valuable tonic on the stomach, and also on the biliary system in general. A ley made of the ashes of the bark of this tree is a good remedy for influenza. The bark of this tree forms an ingredient in Foreman's anti-bilious pills.

COLUMBO ROOT.--(*Frasera Verticillata.*)
[CAH-HUH-SKEE.]

The American Columbo is a native of the United States and grows in great abundance in many parts of the Southern and Western States. It is a stately elegant plant, and has various names, such as Columbia, Indian Lettuce, Meadow-pride, Marietta Columbo, and wild Columbo. Its root is triennial, which means lasting three years; it is rough, spindle-shaped, yellow, running horizontal in the ground, sometimes to the length of two feet. It is one of the most grand and stately looking plants in the American

forest--its stem is large and rises from five to ten feet high it is nearly square and furrowed at the sides, it sends off its leaves, which are of a deep green color, at intervals of eight or ten inches. Some of the leaves are radical, forming a star, spreading out on the ground, the remainder of them grow in whorls around the stem from four to eight to a whorl, each whorl as it is nigher the top, contains smaller leaves, its branches are few, except near the top, where they form a handsome pyramid, crowned with numerous flowers of a yellowish, white or cream color.

The root is the part used, and in its fresh state or when it is first dug, is both cathartic and emetic, but when dry it is an excellent bitter tonic, acting as a powerful strengthener to the stomach in dyspsy or indigestion, and in dysentery or looseness of the bowels, arrising from a superabundance of bile; the proper mode of using it is in powders. The powders taken in cold water, will generally check vomiting or puking, and always will be found beneficial in cholics or cramps of the stomach, want of appetite, &c. The pulverized root steeped in good whiskey, with the addition of a little peppermint taken three or four times a day in table-spoonful doses, will, in most instances, moderate the puking which so often occurs with pregnant women. It is a valuable medicine in fevers and bilious cholics in the last stages. It is antiseptic, that is, it prevents and removes putrifaction; for this purpose, it should be taken in a decoction internally and applied externally in poultice. The tincture is a valuable and safe family medicine, useful to strengthen the digestive organs and increase the appetite.

DOG-WOOD.--(*Cornus Florida.*)

Dog-Wood is found in almost every part of the United States, and are so well known as to render a description unnecessary. Dog-wood is a tonic, stimulant, antiseptic and astringent. It is valuable in all cases of intermittent fevers, by which is meant all fevers that go off and return again; and the only reason why it cannot be given in other fevers, is, that when given in actual fever, it increases the pulse; hence you will see the necessity of never giving it except when the fever is entirely off. It is greatly superior to the peruvian bark obtained in the shops in all cases where the peruvian bark would be applicable. When it pain or griping of the bowels, a few drops of laudanum given in the bark will obviate the difficulty. The bark of the root is the strongest, and the next in strength is the bark of the body and smaller branches. The bark should be taken from the root or tree and cleansed of all dirt and well dried before it is used, as it is less apt to affect the bowels than when taken in a fresh state. The best mode of administering it is in powders, dose is from twenty-five to thirty-five grains. The flowers in tea or decoction, or in spirits is good for cholic. The ripe berries of the dog wood, digested in good spirits of any kind, make an excellent bitter for common purposes and are well adapted to persons of weak stomachs. An excellent family bitter may be made as follows: take equal quantities of dogwood bark, yellow bark and sarsaparilla root, digest them in common spirits until the strength is extracted; this constitutes an excellent morning bitter for family use. The dog-wood bark boiled to a strong decoction, forms an excellent tonic both for persons of weakly habit, particularly infants, that have had their health impaired by long continued bowel complaints or chills and fevers, &c. The bark boiled to a strong decoction and thickened with wheat bran, rye or corn meal, forms a valuable poultice to reduce swellings, allay inflammations, &c. The internal use of the dog-wood, always renders the pulse quicker, and in some instances fuller than it naturally is.

STAR-ROOT.--(Aletris Abba.)
[OO-WA-SCO-YOU-HU.]

Star-Root, sometimes called Unicorn or Blazing-Star grows in low lands or hill sides, and often on very poor land. It has a rough, wrinkled, perennial root; the caudex or main root is about the thickness of the little finger, and the lower end often dead or rotten, from the main root issues many small blackish fibres, the whole root is of a dirty dark color and full of little holes. The leaves are radical, which means implanted in the ground or putting out from the root without any stem or stalk they are a pale evergreen, and in the winter lie flat on the ground, they are smooth and spear-shaped. The scapa or flower stalk rises from eight to eighteen inches high, upright, naked and terminates in a most beautiful spike or tassel of small white flowers.

Star-Root is a valuable tonic and general strengthener of the system, it is very bitter, though not unpleasant. It is excellent for woman in child-bed (puerperal) fever, after the tsomach and bowels have been emptied with the proper medicines in this disease, it should be given in a decoction lukewarm, two or three tea-cupsful a day. It is a great strengthener of the stomach and womb, and assists in casting off the morbid matter from the womb. This root is one of the best articles in the Indian Materia Medica, to prevent abortion, and is earnestly recommended for the constant use of pregnant women that are subject to miscarry: the best mode of taking it in this case, is in spirits. It is a very useful commonplace bitter; it is an excellent medicine when combined with other articles for suppressed menstruation, especially where the general health is impaired and a tonic or strengthener needed.

Star-Root is good for coughs, consumptions and diseases of the lungs, as it not only strengthens the general system but also promotes expectoration and perspiration.--Dose, half a teaspoonful of the powdered root morning and night. It sometimes, and not unfrequently, produces soreness of the mouth; on the first appearance of this, its use should be discontinued for a time, and some other expectorant employed in its stead; on the disappearance of those symptoms, its use may be resumed. It is said by some to be good in the treatment of rheumatism, stranguary, jaundice and flatulent cholic.

MUSTARD--WHITE AND BLACK.
(Sinapis Nigra et Alba.)

Those plants need no description, being cultivated in almost every garden in the Union for culinary purposes.--The black Mustard is stronger than the white. The ground seeds are much used at table to increase the appetite; for this purpose it answers admirably in phlegmatic or inactive stomachs. A table-spoonful of the ground seeds taken on an empty stomach, will sometimes operate as an emetic, repeated doses of the unbruized seeds, will operate as a mild laxative, but its principal virtues reside in its tonic and stimulating properties. Digest the bruised seeds in wine, and they constitute an excellent tonic in fever and ague, nervous fever, dropsy, palsy, &c.

They also form a principal ingredient in the beer for palsy. The bruised seeds taken every morning, has of itself cured phthisic or asthma. In sinopisms it is good applied to the soles of the feet, to raise the pulse, and produce a revulsion from the head. To prepare this sinopism, or plaster, take the ground or bruised seeds, wet them with vineegar and spread them on cloths, moisten the skin also with vinegar and confine the plaster on the part: it may be applied by making paste and spreading it on cloths, and sprinkling the ground mustard-seeds over the paste, and apply it as above directed, after the skin has been moistened with vinegar or spirits. These plasters are excellent in all acute diseases, where the circulation is languid, and the extremities become cold.

HORSE RADISH--(Cochleari Armoracea.)

Horse radish is a garden herb, and is common in every part of the country. It is a stimulant tonic, and diurretic. As a diurretic, it is useful in gravel, and may be taken in decoction or digested in spirits; the root sliced and steeped in vinegar, and used as a condiment with meat, is good to provoke the appetite, and is good to persons of sedentary habits, and weak digestive powers. The root steeped and applied externally, acts powerfully as a local stimulant, and is good applied to joints affected with rheumatisms.--Applied to the bowels and feet in typhus fevers, and accute diseases, it is equal to the mustard seed. A syrup made of the root is useful in phthisic, (asthma,) and bad colds, after the inflammatory stage ceases. Taken in decoction or in spirits, it is good for obstructed menses. It is a valuable article in palsy, particularly where the disease is in the tongue and mouth; for this purpose, chew the roots.--I should have said that the root is the only part used as a medicine.

BALM--(Melessa Officialalis.

This herb is too well known to require a description.--It is gently stimulant and tonic. It is excellent in old colds taken night and morning, sweetened with honey; the addition of a little vinegar will render it much better, and

somewhat more palatable. It is valuable in typhus or nervous fevers; after the stomach and bowels are prepared for its use. It is also valuable in chills and fevers; for this purpose drink a large quantity of the tea as warm as it can be taken, on the approach of the chill. Balm is a very good family medicine, and is quite harmless in its effects.

YELLOW POPLAR--(Lirioderadran Tulipifera.)

This noble and beautiful tree is a native of the American forest, and is so generally found in all parts of the country, as to render a description entirely unnecessary.--It is sometimes called *white root, American Poplar, tulip tree;* this last name was given it from a fancied resemblance between its blossoms and those of the tulip. The bark of the root, trunk and branches of this tree has been esteemed by the Indians as a most valuable medicine; this opinion is now sustained by many of the most distinguished physicians among the whites, both in the United States, and in Europe.

The bark of the root is the most active, and is consequently preferred to that of the trunk or branches. It is a valuable bitter tonic, gently laxative, combined with the dog wood bark, it is equal, if not superior to the American bark, and how very strange it must appear to every reflecting person to see those affected with disease, paying high prices for foreign medicines, the strength of which must be diminished by age, and many times adulterated with other substances, wholly inapplicable to the diseases for which they are intended. while their own farms abound with an article equally good, if not superior to the foreign article, even if it could be obtained pure, and in a fresh state. The poplar bark is valuable in dyspepsia, in dysentery, and in chronic rheumatism. If given in acute rheumatism, where is also some inflammatory fever, it will increase the fever, and I here repeat, and hope it will be remembered by the reader, that stimulants should never be given in fever which continues without intermission; but they may always be given with advantage, and safety where there are periodical cessations of fever, by which is meant such fevers as cool off once in twenty-four or forty-eight hours. It is anthelmintic or a good worm medicine; the best mode of administering it for worms is in powder combined with honey. Children that are subject to warm spasms may be entirely relieved by taking a tea-spoonful of the powder in honey every morning on a fasting stomach. It is good in *cholera infantum,* (puking and purging) among children after the stomach has been cleansed, or the puking checked by the use of the cholera morbus root. The pulverized bark digested in whiskey forms an excellent family bitter; giving a tone to the stomach and bowels, especially when the bowels are in a relaxed state, which require strengthening medicines.

Several physicians among the whites say that in their hands it has entirely relieved breast complaints attended with symptoms similar to those of pulmonary consumption, where the patient had hectic fever attended with night sweats, weak bowels, &c. They administer the powder combined with laudanum. For women afflicted with hysterics and weakness it is an excellent medicine. It may be given in decoction, tincture or powder, but in most cases the powder is the best, except where a family bitter is desired. The bark should be taken from the tree in the month of January or February, and dried; as soon as dry it should be pulverized and bottled for use.

YELLOW SARSAPPARILLA. (Menispermum Canadansis.) [TSU-YOU-E-YOU-STEE.]

Yellow Sarsapparilla is a native of the United States and grows mostly in rich moist lands, in river and creek bottoms. It has a long, yellow, woody, perennial root, with but few fibres, the root runs very shallow, and is very easily pulled from the earth. Its vine is woody, small, of a dark green or brown color, running from six to twelve feet high, turning around whatever happens to be near it. Its leaves are few and scattering, deeply indented, and in shape bears a strong resemblance of the maple leaf.

The root is the part used as medicine. It is valuable in all diseases of the skin; it is a good laxative bitter tonic, useful in debility, giving tone to the stomach, and vigor to the nervous system. Where the system has been injured by the use of mercury and is laboring under great debility, it certainly is a valuable medicine. It has a gentle tendency. to determine the fluids to the surface, or excite prespiration. It is good for weakly females, afflicted with weak stomachs and bowels; it is also good in the treatment of venerial--it may be used in the decoction or bitters.

WHITE SARSAPARILLA--(Smilax Sarsaparilla.) [AH-SO-E-NO-O-ONA-KER.]

This vine is a native of the United States, and also of the Spanish West Indies. It is a small running vine, of a dark color outside, and a pale white within, the main vine is about the size of a common goose quill, it bears a strong resemblance to the yellow Sarsaparilla, and possesses similar and medical qualities, though in a more active degree; it is more bitter to the taste than the yellow, the leaves are not quite so large, and of a darker green, the root when broken is much whiter than that of the yellow--it does not make quite so pleasant a bitter for common family use as the yellow does. The white sarsaparilla grows mostly in rich cultivated lands, and along the borders of meadows. I have used the white Sarsaparilla in bitters with great success in nervous debility. It is also useful in dropsy, gout, scrofulous sores, diseases, and diseases induced by the use of mercury. For that loathsome and disgraceful disease pox, it is an excellent remedy: in this disease it is used in decoction in combination with the yellow sarsaparilla and wild mercury: it is also a great tonic and strengthener of the digestive organs. It is valuable in dyspepsia, and chronic affections of the liver. When it is taken in chronic diseases, or where there is no fever and much debility, it should be taken in spirits as bitters. But when used where there is fever, as in pox, &c., it should be taken in decoction, made by cutting or splitting an ounce of the root fine, boil it in a half gallon of water, down to a quart, of this, drink from a half pint to a pint a day, or in larger quantities if desired; for, although it possesses great power; yet it is entirely innocent in its operation on the system.

WILD HOARHOUND--(Eupatorium Pilorum.)

The wild hoarhound is too well known to need a description.

The leaves are remarkably bitter to the taste. The leaves are the part used, and are a valuable laxative bitter tonic. They should be boiled to a very strong decoction and sweetened with honey to the consistency of syrup, taken in table-spoonful doses, three times a day. It is good in consumption and breast complaints generally; also in colds, phthisics, &c. It may be used in all cases where laxative tonics are required, particular in chronic diseases, in advanced stages where the stomach requires a bitter tonic. It produces a disposition to sweat, and gently increase the secretion of urine, these added to its tonic and cathartic properties render in a valuable medicine.

SNAKE HEAD--(Chelcna Glabra.)

This plant has a perennial root, or one that is not killed by the frosts of winter, its stem is square sometimes erect but often bendnig. The flowers grow out at the end of the stem, and of different colors in the different varieties of this plant, as white-spotted, white red and purplish; the flowers in shape resembles the head of a snake, with its mouth open; the leaves are opposite of a dark green color, and bear a slight resemblance to mint leaves, they turn black on being dried, and are very bitter to the taste. The leaves are the best part for the medicinal purposes, and may be used in decoction, in powder, or tinctured in wine or peach brandy.

It is a powerful bitter tonic, and acts powerfully on the digestive organs, it increases the appetite equal to any I have ever administered. Such persons as are afflicted with biles, and sores or eruptions of the skin, will derive great advantage from its use. It is also useful in fevers when a tonic is required. The powders taken in large doses is cathartic, and in some instances acts as well as a warm medicine; but should not be given when there is much excitement or fever. In worm complaints when there is but little fever it may be used to advantage.

ANGELLICA--(Commonly called Angellico.)

This plant is well known and grows most plentifully on rich hill sides, and mountainous countries. The root is possessed of the strongest medicinal virtues, but the whole plant partakes of the same, though in a less active degree. Combined with dogwood bark and yellow poplar root bark, it is a good tonic after long spells of ague and fever, it may be taken in spirits if preferred. The decoction sweetened with honey, taken at bedtime, say a gill, is good for colds of long standing, obstructed menses, &c. It is an excellent stimulating and sweating medicine, and is peculiarly adapted to weakly females, of nervous and phlegmatic habits. Either in decoction or tincture it is good for flatulent colics, and when tinctured is quite pleasant. A strong decoction of the root makes a very good gargle for soar throat, and mouth.

SOLOMON'S SEAL--(Convallaria Multiflora.)
[OO-TE-TI-SI-KEE.]

The leaves of this plant are of a dark green color, ribbed, clasping the stem, and of an oblong or oval shape. The flowers grow out along the side of the stalk, and forms a kind of angle with the leaves.

The root is the part used. It is a mild tonic, and is useful in general debility, and diseases of the breast or lungs. It is also good for weakly females afflicted with whites or profuse menstruation, it may be used either in tea or syrup. In dysentery or old bowel complaints it is an excellent remedy, and seldom fails to effect a cure if its use is persevered in for any length of time.

GOLDEN SEAL--(Hydrastus Canadensis.)
[TAH-LOA-NE-GA-OO-NAH-STA-TSE.

Golden seal has a perennial root, or one that is not killed by the frosts of winter, it is of a bright yellow color, the main root is crooked, rough, and very knotty, with many small roots or fibers. Its stem rises from ten to fifteen inches high, round, straight, and commonly bears on the top two leaves, they are rough, and bear some resemblance to the maple leaf. It produces but one flower which is succeeded by a beautiful red, fleshy berry, which contains the seeds, It is a valuable bitter tonic, and may be used in all cases of general debility, as it will strengthen the digestive organs, improve the appetite, and in crease the tone and strength of the organic system throughout, it may be used with great advantage when recovering from fevers or other diseases which cause debility, it is useful in relieving the disagreeable sensation arising from indigestible food, which is so often experienced by those laboring under dyspepsia. The dose is about a tea-spoonful of the pulverized root, infused in hot water. It may be used alone or combined with other tonics. The decoction of this root, used externally as a wash or bath, is good to allay local inflammations.

FENNEL, SWEET FENNEL.
(Anethum Foenicular.

Sweet-fennel is a garden herb, and is too well known to require a description. The seeds of the fennel are a pleasant aromatic tonic, pulverized and sweetened with honey or sugar, or in decoction sweetened; they are admirably adapted to pains in the stomach and bowels, colics, &c. There are few better articles for young children afflicted with flatulent colic, than sweet fennel seeds.--Given to women in labor, when the pains are short, followed by sickness at the stomach, they will generally produce good effects by relieving the sickness, and strengteening the system, so as to enable nature to perform her task. The fennel seeds may be used in bitters either alone or with other articles as they will greatly improve the taste of other tonics. The oil obtained from the fennel seeds is valuable for colds, colics, &c.

WILLOW.--(Salix.)

There are several varieties of the Willow, all possessing similar properties as medicine, the white is some the strongest or most active.

The bark of the Willow is tonic, and may be employed as a substitute for dog-wood or Peruvian bark. It is generally taken in decoction, say half a gill three or four times a day. But the principal use made of it by us is in poultices, made by thickening wheat bran or rye meal in a strong decoction of the root or bark of the root.

(The English name not known.)
[OO-NA-KER-OO-NAH-STA-TSE.]

This is a small white tender looking root, never growing larger than a common pea and seldom so large, when the stalk is taken up several small balls or roots are found under it very much resembling a hill of yam-potatoes, only so very much smaller; the external appearance of the root bears a strong resemblance to the artichoke, and when broken it looks clear like the artichoke, and possesses a taste not very dissimilar to the taste of that root; these little balls or roots are generally round, but sometimes inclined to be long, never exceeding a half inch in length to the best of my knowledge; these little knots or roots are attached by a small fiber which extends from the main root and then from one to the other. The stem grows from six to twelve inches high, small, smooth and divides into three branches, sometimes only two, near the top; each branch has three smooth leaves, oval and scollaped, or indented irregularly at the outer or extreme end of a light or pale green color, seldom if ever more than an inch in length and narrowed at the end which is attached to the stem. The stem is of a whitish purple color and not thicker than a course sewing needle. The stalk and leaves of this plant, bear such a strong resemblance to that of the cholera morbus weed, that the one is often mistaken for the other until the root is examined, which bears no likeness whatever. The root of this plant is a valuable tonic. Persons that have become lean and emaciated, have often recovered their flesh by the use of this article alone. Infants when very young, appearing to dwindle and pine away, will derive great benefit from the use of this root. The manner of using it is to bruise it and put it in cold water and make it a constant drink. Women that have been married for a number of years and had no children, on making a constant drink of this root, have been blessed with a healthy offspring Where the general health appears to be good, I believe this root will in most instances prove a cure for barrenness.

WILD CHERRY-TREE.--Prunus Cerasus.
[TEX-TAH-YAH.]

The bark of the wild cherry tree is tonic and astringent;

as a tonic it ranks next to the dog-wood bark, and may be combined with that article, with great advantage. It may be given either powdered in substance as the other barks are, or it may be given in decoction; a handful of the inner bark, to a quart of water, taken in teacupful doses three or four times a day. It is a good tonic in intermittent fevers, and in bilious fevers in the advanced stages, when tonics are requisite, the cherry bark, in wine or French brandy is a most excellent tonic, particularly where the stomach and bowels are debilitated. Like other tonics it should never be taken when the fever is on. The gum of the wild-cherry tree is equal to the gum arabic obtained in the shops, and may be used in all cases which call for the arabic gum. The bark of the tame cherry tree of this country, digested in spirits makes a wholesome, and tolerable pleasant family bitter.

A strong decoction of the wild cherry tree bark is valuable in the treatment of jaundice, as may be seen by turning to the treatment of jaundice, the inner bark may be bruised and taken in spirits if preferred. The bark of the root in decoction forms a valuable wash for old sores, and foul, ill conditioned ulcers.

BLACK HAW.
[CAH-HE-CAH.]

This shrub or bush grows in many parts of the United States; it bears a small fruit which is considered by some very delicious; this bush is so well known in the country where it grows as to render a description needless.

The bark of the root of this bush is tonic and diaphoretic, combined with dog-wood, or wild cherry tree bark, it forms a good tonic in intermittent fever or ague and fever. The bark of the root in spirits, is a very good family bitter. When it is necessary for pregnant women to take a sweat, this-Haw-root bark is used combined with other articles.

CHOLERA-MORBUS ROOT
[SAH-KO-TSE-KEE.]

The herb is mostly found in bottoms and on the banks of the streams in shady places. It has a whitish, fibrous root, rather small, smooth, growing from six inches to a foot high--leaves smooth, roundish with an indentation on each side, of a bright green color. The top of this bears a strong resemblance to the top of the Oo-ne-kee-oo-nah-ste-tse. It generally grows some taller than that herb, and the leaves are of a brighter green. When dug they are easily distinguished; one having a fibrous root, and the other small balls or lumps attached by fine thread like roots. The root is the part used; it is tonic, anticeptic, antiemetic; and is a certain remedy for cholera-morbus.

IRON FILINGS.

Commonly called steel or iron dust, is made by heating a piece of iron or steel to a very great heat, and rubbing it with rolls of brimstone, and let the melted parts drop into a vessel of water; then reduce it to a fine powder; and sift it through a muslin cloth; it may be given in doses of from 8 to 20 grains to suit the diferent age and strength of the patient. This is a most valuable tonic, good in dropsies, liver complaints, weak stomach and bowels, and in most cases of debility.

GENTAIN.

This herb grows mostly in dry, oak and hickory land. It has a long round, tapering, perennial root. Sometimes of a light and other times of a darkish brown color. Stems are many, erect, round growing from two to three feet high. Leaves are opposite, lower ones connate or joined together, so as to have the appearance of being but one, with the stem passing through the centre. Flowers grow at the base of the leaves, of a reddish color, and are succeeeded by large, yellow berries crowned with four or five leaflets, which are the calyx of the flower. Flowers are from two to six in number. The root is the part used; it has a pleasant bitter taste; it is tonic, stimulant and cathartic; it is one of the best laxative bitter tonics in the Indian materia medica. It is one of the most valuable remedies for weak stomach and hysterical affections; for this purpose, it may be taken in spirits or bitters. It may also be taken infused in doses of a tea-cupful three or four times a day. It is an excellent medicine for dyspesy, either alone or combined with other articles--it prevents the food from souring and oppressing the stomach in short it is one of the best common place medicines in my knowledge.

CLASS No. V.

ASTRINGENTS.

Astringents are medicines that are used to render the solids more dense and firm, in order to correct debility and looseness. They exercise a very powerful and extensive influence on the system and are of greater or less utility in the treatment of most diseases which the human family are subject. In the incipient or forming stages of diseases, this class of medicines, if properly administered, will often throw it off entirely. Their free use when recovering from disease, has a tendency to prevent relapses. Medicines of this class must be used sparingly, or omitted altogether in some cases, such as obstinate costiveness, high fever, attended with extreme dryness of the mouth, &c. Astringent tonics, are such as relieve floodings and hemorrhages of every kind and may be advantageously employed in all profuse evacuations and relaxed states of the system.

NEVER WET.
[CAH-NA-SEE.]

This valuable and singular plant is found growing in the water, particulary in slow running spring branches, in the Southern parts of the United States. The stem always grows to the surface of the water let it be what depth it may, before its leaf comes out; the leaf always lies on the surface of the water, it is from six to twelve inches long and two or three inches wide: its color is a pale or light green: its surface is remarkably smooth and glossy as if covered over with oil, so that the water will not wet it--from this circumstance it takes its name Never Wet. The whole leaf is very tender, thick and fleshy. They are excellent, wilted or scalded and spread on burns of any kind. They may be bruised or beaten and applied in the form of a poultice; they are also a good dressing for blisters and ulcers, or sores of almost every kind, giving much relief to the pain so generally experienced in scalds burns, sores and other inflammations.

BALM OF GILEAD.--(Amyris, Gildensis.)

The tree known in this country by this name, is mostly cultivated as an ornament for yards. The genuine Balm of Gilead is a native of Asia, and grows near the city of Mecca, on the Asiatic side of the Red Sea. That growing in Gilead, was anciently esteemed the best, and was thot by the ancients to possess remedial virtues for almost every disease--hence this tree received the name of the *Balm of Gilead.*

The American Gilead is a species of this tree, and as I before told you, is mostly cultivated as a yard ornament in the Southern and middle States, it cannot bear the severity of Northern winters. Its leaves are large, smooth and beautiful, nearly of a heart shape. The bark of the young tree is smooth, both the bark and leaves resemble those of the lumbardy poplar, but it does not grow so tall and erect.

The tincture of the buds is good for cholic, old bowel complaints, both among children and grown persons; it is also good for chronic rheumatism and may be rendered better for the rheumatism by adding the bark of prickly ash. For rheumatism, the tincture of the buds must be applied externally to the affected part, and a tincture of prickly ash bark and the Gilead buds drank as bitters, say three times a day. This tincture is also good for old venereal complaints; steeped in water or taken in tincture. It is excellent for persons of weakly phlegmatic habits.--The bark and leaves possess medicinal virtues, but in a less active degree than the buds. The buds stewed in deer or sheep suet, makes a most excellent salve, when combined

in proper proportions, but if too strong with the buds, it will irritate the wound and make it worse. The buds are valuable in salves or ointments, for tetter-worm, scald head, burns, &c. The tincture of the buds, or gum or rosin off of the buds put into the hollow of an aching tooth, will generally give relief. The buds are found on the tree nearly all the year, they are large of a brownish color, and contain a considerable quantity of a kind of a gum-rosin or balsam. A syrup made of the buds and sweetened with honey is an excellent wash for the sore mouth.

THE OAK.--(Quercus.)
[GEE-GAH-GAY-AH-TAH-YAH.]

We have several species of the Oak, as the black, white, red, &c., all possessing similar medical qualities. The bark is the part used, and is astringent, ionic, and antiseptic.

After long fevers, intermittents, indigestion, chronic dysentery, or any debility of the system, it is a most valuable astringent tonic--in decoction is the best mode of using it. It constitutes the best bathe in my knowledge for persons of weak, debilitated or relaxed habits. Repeated instances have occurred, in which persons, especially children, have been reduced to mere skeletons, by long continued bowel complaints, their stomachs had become so irritable as to render it impossible to relieve them by medicines taken internally, and were restored to health by bathing in a strong decoction of the Oak bark twice a day. The decoction thickened and applied as a poultice, is good to reduce inflammation and prevent mortification. I believe the red oak to be the best for poultices. Children afflicted with chills and fevers, when too young to take tonics into the stomach; have often been relieved by bathing them in a strong decoction of dog-wood and red oak bark at the time when the fever was off, and applying the pulverized bark of the dog-wood to the waist, wrists and ankles, by the means of bandages--flannel is the best: the bath should be administered about six or seven hours before the chill is expected, and as soon as the patient is taken from the bath, the above application should be made, and it should remain until the approach of the chill.

DEWBERRY.--(Robus Procumbens.)
[OO-TO-SEE-NER-TUH.]

The root of the Dewberry brier is astringent and tonic, and are valuable in the treatment of venereal. A decoction of this root and the root of the blue flag, has often cured this dirty complaint in a few days. The root boiled in new milk or water, is good for persons afflicted with chronic, or old bowel complainsts, particularly in aged persons, and such as have weak or debilitated constitutions. The tincture is good for persons of weakly phlegmatic habits, perhaps better than the decoction. The decoction sweetened with honey, makes a healing wash for sore throat and mouth, and will sometimes cure the thrush.

The Blackberry possesses the same medicinal properties that the Dewberry does, but in a less active degree.

SWEET-GUM TREE.
[TSE-LAH-LEE.]

The Sweet-Gum is too well known to need a description: it grows in great abundance in many places in the U. States, generally in rich bottoms or low lands. The inner bark, leaves and rosin, or gum, are the parts used for medicine: the rosin or inner bark is excellent for diarrhoea, dysentery or flux. When the bark is used, it should be boiled to a strong decoction in new milk or water, and the decoction taken in tea-cupful doses every hour until relief is obtained. It may perhaps be necessary to cleanse the stomach and bowels with a cooling cathartic previous to taking the Sweet Gum tea: the rosin is valuable in bowel complaints, but must be used with caution, lest it should prove too binding, and thereby produce too much excitement in the system. The leaves bruised and steeped in cold water, is a good wash for scald head: the gum or rosin, forms an excellent ingredient in salve for wounds, sores, ulcers, &c. It is also good mixed with sheep or cow's tallow for itch: it seldom fails to cure the itch if applied perseveringly.

PERSIMMON.

The Persimmon grows in most parts of the Union and is well known to almost every person.

The bark of the tree and root and unripe fruit are highly astringent. The bark of the root forms an ingredient in the decoction or beer for venereal, as may be seen in the treatment of that disease. A strong decoction of the inner bark sweetened with honey is a valuable remedy for sore throat and mouth. Made into syrup, it is good for thrush, and may be made better by mixing with it a liitle finely pulverized Borax. The decoction is a good astringent wash in all cases where astringents are required, used as a wash or bath to the fundament it is an excellent remedy in case of piles; it may be applied by wetting lint in the decoction and applying it to the fundament.

COMFREY.--(Consolida.)
[OO-STER-OO-STE-LUR-E-STEE]

Of this plant there are two kinds, the wild and garden Comfrey. Of the two species the garden comfrey is some the best, owing to its containing more mucilage or jelly, and not being quite so hard and tough as the wild.

A handful of the roots boiled in new milk and drank freely, is good for flooding after child-birth. A gill of the milk in which comfrey root has been boiled, given every half hour, is amongst the best remedies for flux or dysentery. The root sliced and steeped in water and used as a common drink is good in clap (gonorrhea,) also for strictures, or heat, in making water it is excellet. The root infused in cold water and made a constant drink, is valuable for pregnant women who are troubled with heart-burn, costiveness, &c. It is also excellent for such females as from sexual weakness are troubled with menstrual discharges and other symptoms of abortion during pregnancy. A poultice made by brusing and boiling the root is valuable to reduce inflammation and prevent mortification. The writer can bear testimony of the efficacy of this poultice, it having removed the inflammation from a wound for him, after it had thrown him into high fevers, without the aid of other remedies. The poultice is made by pounding or bruising the root fine, boiling it in new milk or water and thickening it with wheat bran or corn meal. The bruised root wet with vinegar is excellent applied to sprains, bruises and bealings; it will often drive back the worst of bealings when other applications fail.

AGRIMONY--STICK-WORT.
(Agrimonia Eupatoria.)

Agrimony has a dark; fiberous, perennial root; its stem is round and hairy, growing from one to two feet high: leaves are alternate, rough, hairy, ragged and unequal, the lower ones the largest. The blossoms grow on a long terminal spike, which is merely a continuation of the main stem--they are of a yellow color; and produce a small bristly brier, which in the fall of the year sticks to clothes that comes in contact with it.

The root reduced to powders and combined with other articles, is much used in the treatment of pox, as may be seen by referring to that head. It forms an ingredient in the nerve powders. A decoction of the root is valuable in habitual diarrhea or looseness, and in all cases of extreme debility.

PRINCES FEATHER--AMARANTH.
(Amaranthus Sanguineous.)

The Prince's Feather is much cultivated in the gardens of this country for its beautiful appearance. It grows from two to four feet high, the whole plant, more or less, exhibits a red appearance, but the bloom is of a beautiful bright red.

The leaves are the part mostly used, and rank among the most active astringents, often relieving floodings and profuse menstruation when other remedies have failed.-- The decoction is the best mode of using it. The quantity taken must be regulated by the effects produced.

CRANE'S BILL.--(Geranium Maculatum.)

This plant is found growing mostly in meadows and low, wet grounds; its root is generally crooked and knotty, blackish on the outside, and reddish within; it has a rough taste, leaving an aromatic flavor behind; its stalks are slender, from six inches to a foot high; bearing seven long, narrow leaves at a joint.

The root is the part mostly used, pounded fine and made into a poultice with cold water, it is the best thing to stop bleeding that I have ever tried. The Cherokees, as well as several other tribes of the Natives, place unbounded confidence in this root as a styptic. It is said by some of the whites to be valuable in profuse menstruation, whites, gleet and obstinate diarrhoe; also, bleeding and hemorrhages of all kinds. I have never tried it any other way than to stop bleeding from a wound, but its promptness and activity, in checking the flow of blood from an artery, induces us to believe, that when properly tried, it will prove a valuable medicine in the class of astringents.

WATER PLANTAIN.--(Alisma Plantago.)

This Plantain is found growing mostly in wet soil, or in the margin of stagnant waters. The root remains through the winter--(perennial.) Its leaves are of a light green color, and very much resemble the common plantain.

A decoction of the root is valuable in all bowel complaints, after a gentle purge has been taken to cleanse the stomach and bowels. But the most important use made of this root is, as an external application to old sores, wounds, bruises, swellings, &c. The roots should be washed clean and then boiled until soft, mashed up and applied in the form of a poultice; the affected part should be bathed in a decoction of the root, before the poultice is applied. This treatment will seldom fail in reducing inflammation, and preventing mortification. It is a most excellent application to old, foul, and ill conditioned ulcers, cleansing them and disposing them to heal.

YARROW.--(Achilla Millefolium.)

Yarrow, both grows wild and is cultivated in gardens; it is so well known as to render a description needless. The leaves are the part used; they are astringent, and will be found good taken in decoction, in gill doses, four or five times a day for hemorrhages, such as spitting blood, bloody piles, bloody urine, immoderate flow of the menses. Also good for bowel complaints, and a weak relaxed state of the system.

CINQUEFOIL.--(Commonly called sinkfield.)

This vine grows in old fields and fence corners, and is something similar to the strawberry. Each stalk bears five leaves, hence it is sometimes called five finger; its blossoms are yellow. The root is astringent, and may be boiled in water or new milk, about a handful to a quart; this decoction is good in fevers and acute diseases when there is great debility: also in dysentery and bowel complaints generally, it sometimes proves beneficial in profuse menstruation.

SKERVISH, FROST-ROOT.
(Erigerom Philadelphicum.

Frost-weed is found in great abundance in the United States; it generally grows in fields, which it sometimes entirely overruns--it is seldom found in the woods. The root is yellowish, composed of many branching fibres; the stem rises from one to three feet high, branched near the top; the leaves are oblong, largest near the ground, becoming smaller as they ascend the stalk or stem: the flowers are numerous, of a yellowish white, sometimes of a purplish blue, and of a downy appearance. This plant continues in bloom until the autumnal frosts, from which circumstance it has derived one of its names, Frost-weed.

The principal use made by us of this plant, is for gravel, and diseases of the urinary organs, but for the further information of the reader, I give the following, which is taken from the writings of different physicians among the whites:

"This plant is astringent, diuretic, and sudorific in a high degree: there are several species of this valuable plant possessing the same medical properties, and indiscriminately used; though distinguished by their botanical or technical names, but not by their common. The medical powers of these plants are very active, and require cautious use. They may be employed fresh or dry, in decoction, infusion, tincture, extract, or oil:--the oil is considered one of the best styptics in medicine. The diseases said to be relieved by this article, are dropsy, suppression of the urine, inflammation of the kidneys, gravel, gout, suppressed menstruation, coughs, hemorrhages, dimness of sight," &c.

I cannot say from personal experience that this article is an infallible remedy for all the above complaints; but I do believe from personal experience, that as active properties

as a medicine, render it worthy the further attention of those engaged in the healing art.

CAT-PAW, OR POLE CAT BUSH.
[TSU-KIR-TAH-LOWL-TEE.]

This bush or tree, is found growing on water courses, such as rivers and large creeks; it grows to about the size of a common peach tree; the bark is of an ash color, variegated with dark spots: it bears a white bloom, which is succeeded by a three, and sometimes a four square pod.--The bark has a very bitter taste, and an offensive smell. It is called by some cotton-wood, by others white wood; it is easily known by the pod:--the bark is the part used.

It is tonic, astringent, antiseptic and expectorant. Taken in bitters, it is good for breast complaints, spitting blood, &c. Boiled to a strong decoction and made into a poultice with rye meal or wheat bran, it forms the best application in my knowledge for white swelling in the first or forming stage.

RASPBERRY.

There are several species of the raspberry, all good for medical purposes. We have been in the habit of using the common black raspberry, but white practitioners prefer the red raspberry:--the leaves are the part used, and are highly astringent. A decoction of them is good for bowel complaint; it is an excellent wash for old and foul sores or ulcers. A strong tea of the red raspberry leaves is said by the whites, to be a most valuable article in regulating the pains of women at or near the period of child-birth.

BAMBOO-BRIER.
[E-TSOG-CAH-NI COH-LER.]

This brier grows in most parts of America: it has a small long vine, full of very sharp spines, the vine and leaves are an ever green, it bears small dark berries, which in the latter part of the fall bear a strong resemblance to the winter grape.

The bark of the root of this brier is astringent and slightly tonic. It is valuable in the treatment of pox, when combined with other articles, as is fully shown in the treatment of this disease. A decoction of the leaves is an excellent wash for scalds, burns, and other foul sores; a salve made by stewing the leaves, or bark of the root in hog's lard with a little beeswak is good for burns, scalds and other sores.

RED-ROOT.
[A-LE-SKAH-LAH.]

This valuable herb grows in great abundance in every part of the Union with which I am acquainted. It grows most abundant in uplands which are tolerably thin, and inclined to be shady, and in such lands as produce pine and hickory, in the Middle and Southern States; but it grows in great plenty north of the Ohio river where there is no pine. Its roots are long and large, covered with a hard, rough, red bark, the whole root, is of a hard woody nature. The top or stem grows from one to two feet high, much branched and crowned with numerous leaves; its blossoms are white, and appear in June and July.

The root is the part used, and is a valuable astringent. A strong decoction of the red-root is my choice wash for cancer; it also forms an ingredient in the decoction for venereal, as is fully shown in the treatment of that complaint. It is excellent in decoction for old bowel complaints, spitting blood, flooding, bloody piles, &c. It is also good for sore mouth and sore throat, in decoction sweetened.

HORN-BEAN--IRON WOOD.
[OO-LI-NER.]

The Iron-wood is a very common growth in most parts of the Union, and is so well known, that any particular description would be needless; it grows to about the size of the dog-wood.

The inner bark is astringent, and forms an ingredient in the decoction for flux.

ALUM-ROOT--(Heuchera Americana.)
[TSE-NO-SCAH.]

This little and useful plant is found growing in most parts of the United States, generally in the woods or forest, seldom in cultivated lands. The stem grows from three to six inches high, of a greyish color. The root is short, and bears some resemblance to puccoon root, not so long, and more of a brownish cast, rough and wrinkly.

The root is the part used, it is very astringent, taken in decoction, or in spirits as bitters; it is very good for old bowel complaints; in decoction, it is good for immoderate flows of the menses; piles, and hermorrhages in general; the pulverized root applied as powders is a most excellent application to malignant ulcers; the decoction made into syrup with honey is good for thrush and other sore mouths. It acts powerfully, as an astringent tonic, and taken in spirits of a morning on a fasting stomach, adding a dram before dinner and supper, it has done wonders in curing dysenteries, after the remedies prescribed by skilful physicians had failed.

WHITE HICKORY.

The inner bark of this tree is a good astringent and detergent, boiled until the strength is extracted, and the decoction reduced over a slow fire, to the consistence of molasses, is one of the best dressings for a cut, (which is not too much inflamed) in the world. When they are ruptured or cut blood vessels, it will, in most instances, stop the bleeding, cleanse and heal the wound. It should be kept in readiness by every family; this may easily be done, by adding to the syrup, a little brandy or proof spirits; it should be applied by dipping lint in the syrup, and binding it to the wound.

KNOT ROOT.

Knot root has a large, woody root, with small roots or fibres issuing from the caudex or head. The stalk and cane bear some resemblance to the rattle weed. It puts up an erect stem, which bears on the top a beautiful tassel of white flowers.

The root is the part used, and is very astringent, reduced to a very fine powder, it is an excellent article to remove proud and fungous flesh.

GREEN SWITCH--YELLOW ROOT.
[TSU-WAH-TO-HAH.]

This bush grows on the banks of streams. The root is yellow; stem rises from three to six feet high, smooth, slim, of a beautiful green color; it is evergreen and in the spring puts forth many greenish blossoms--leaves painted, and slightly indented. It is an astringent and tonic; a tea or decoction of the root is good for the piles. The ashes burnt from the green switch is an excellent application to cancer.

CLASS No. VI.

SUDORIFICS AND DIAPHORETICS.

Diaphoretics and Sudorifics are medicines that promote prespiration, strengthen the living power, and give firmness to the muscular fibres.

Sudorifics are such as produce copious sweating.

Diaphoretics are such as produce only gentle perspiration or moisture of the skin, but they will both be placed in one class.

SENEKA SNAKE ROOT--(Polygola Senega.)
[OOYER-LEG.]

The stalk of this plant grows about a foot high, upright and branched; its leaves are somewhat oval and pointed; flowers white; the root is variously bent and twisted, rough and of a jointy appearance, thought to bear some resemblance to the tail of a rattle snake; hence it is sometimes called rattlesnake root; several opinions have been given

as to how it first received the name, seneka snake root.--Some say, it was called, for the tribe of Indians who first used it, as medicine; others that it obtained its name from its efficacy, in the cure of the bite of the snake.

The different tribes of Indians have long ascribed to this root medical proprieties of the most active and important kind; it is thought by them equal, if it does not surpass any root in the American forest, for its various and useful effects on the human system, as a medicine. This opinion has been sustained by many respectable physicians among the whites at Percival, Millman, Chapman, Tenant, Archer and others. It is sudorific; diuretic, emmenagogue and cathartic.

It is certainly one of the most valuable remedies in the world for obstructed menses. It may be taken in decoction or combined with other articles, and used as bitters--if when taken in decoction, it should produce sickness or vomiting which is sometimes the case, when the stomach is weak and irritable--add to the decoction, a little angellica, calimus or ginger.

It is good in colds, pleurisy, acute rheumatism, and inflamatory complaints. In all dropsical swellings, it is an excellent article, as it increases the tone and strength of the urinary organs, while at the same time, its laxative properties keep the bowels in a proper state.

For colds, croup, and menstrual obstructions, it should be taken in moderate doses, often repeated, until the desired effect is produced. It may be given with safety and advantage in all cases, where a sweat is required, after the stomach and bowels are prepared for it, except to females in a state of pregnancy, in this case, it should never be used, on such as are subject to immoderate flow of the menses.

This root forms an ingredient in Dr. Wright's famous beer for consumption. We use it in the Chalybeate pill.

The root should be pounded or pulverized, as it is very slow to yield its strength.

Much has been written with regard to its virtue in the cure of the bite of the snake; we have never used it for this purpose, believing that the remedies prescribed for the treatment of animal poisons are superior to this root, but should a case occur where this root was at hand, and the remedies prescribed under that head could not be obtained,

we would give it a fair trial; the mode of using it is internally in tea or decoction, and externally, to the wound

INDIAN CUP-PLANT.
[OO-TAH-NER-CAH-NE-QUAH-LE-SKEE.]

This plant has a large, long, crooked perennial root, and forms a joint where the old stalk grew, which leaves a hole in the root where it decays, and from each of these joints issue fibers. The stalk is square, with the sides concave, which makes the corners very sharp, and grows from six to eight feet high--leaves are very large, grow opposite and are indented on the edges with large deep teeth, they are united at the stalk or base with the edges so raised as to form a cup, which would contain a spoonful or two of water, and found growing mostly in rich bottom lands.

The root is the part used, and requires long steeping to extract the strength. It is one of the most valuable articles in the Indian Materia Medica to promote perspiration and give vigor to the living power; it is tonic, and is an excellent remedy for weakness, inward bruises, &c. For ague and fever and bilious fevers in the last stages where tonics are needed there is nothing better. It is the most efficient medicine in our knowledge to dissolve and carry off ague-cakes.

PENNYROYAL.--(Hedeoma Pulegiorides.)
[COH-WAH-SUR-GEE-SU-SLE.]

This plant grows in great abundance in every part of the country, and is so well known that I need not describe it. A decoction of this plant is a warm stimulant and diaphoretic. It is expectorant, the expressed juice sweetened with honey or sugar is useful in colds, coughs, particularly whooping cough. The tea drank freely just before going to bed is a valuable article for obstructed menses, its use should be continued until relief is obtained. Pennyroyal tea is often used to advantage as a drink, in promoting the operation of emetics. A free use of the decoction at the commencement of fever, will often throw it off and give entire relief. The essence of Pennyroyal is valuable in all cases in which I have recommended the decoction or tea.

The herb should be gathered just before, or about the time it blooms, tied up in small bunches or bundles and hung up where it will keep dry; it makes one of the most pleasant and useful teas in common light family sickness with which I am acquainted.

SPICEWOOD.--(Laurus Benzoin.)
[NA-TAH-TLAH.]

The Spicewood is found in most parts of America, and is so generally and so well known, as not to need a description. It is generally found in rich, uncultivated, marshy places, about the edges of branches and ponds.

A decoction of the twigs, bark or root is a good diaphoretic, and may be used to advantage in colds, coughs, phthisics, croup, &c., taken in gill or half pint doses, every hour or two, it is good in female obstructions; in this case it should be taken occasionally as much as the stomach will bear, and the feet bathed on going to bed. A strong decoction of the bark or root sweetened to syrup and given to a child on the first symptoms of croup will generally give speedy relief. The berries boiled in bark is good for dysentery and bowel complaints and it some times expels worms. But all has not been told yet about this useful shrub; it is a most valuable article in the treatment of white-swelling, although the whites think themselves perfectly acquainted with the medical properties of this bush, yet they appear entirely ignorant of this important medical virtue; for the manner of using it in white-swelling, refer to the treatment of that disease.

RATTLE-WEED.--(Botrophis Serpentaria..)
[OO-LE-LAH-STEE.]

This berb is called by different names, such as squaw-root, squaw-weed, &c.; it is found in every part of the Union, growing mostly on the sides of rich hills, and mountains, also in rich bottom wood-lands. The most common name in the western country is squaw-root, I believe, and is said to derive it from the extensive use the Americans saw the Indian women make of it in the settling of America. The stalks grow from two to six feet high, nearly round, smooth, and branched at top; it bears a kind of tassel or bunch of berries, which when ripe on being shook, makes a dry shattering noise, and from this fact it is called rattle-weed.

The root is the part used, it is sudorific, tonic, diuretic, anodyne, emmenagogue, and slightly astringent. It may be given either in powders, decoction or tincture. In decoction it is valuable for colds, and female obstructions, when the obstruction is of long standing, and the general health impaired by it, this root should be used in tincture or bitters, and may be advantageously combined with other articles. It is valuable in coughs and consumptions. Obstinate bowel complaints have been speedily relieved by drinking a decoction of this root. It may be used in bitters, combined with spikenard, in the latter stages of pregnancy to great advantage, but its use should not be commenced before the end of the seventh month. It is an excellent article in the treatment of rheumatism, in acute or inflammatory rheumatism it should be taken in decoction; for chronic rheumatism in tincture or bitters it is much the best.

HEART LEAVES.
[LE-SQUAW-CLA.]

The top, root and blossoms of this well known plant all possess medical virtues. A tea of the leaves, roots or blossoms, taken a pint morning, noon and night, is useful to relieve hysterical or nervous debility, and strengthen women of sexual weakness. This tea is excellent for girls whose periodical evacuations are not regularly established, and for women whose courses are about to leave them from their age, according to the laws of nature.

The strong tea taken in large quantities and as often as the stomach will bear is good in typhus fever, and chronic cases of ague and fever, it should be commenced just before the chill is expected, and continued until perspiration or sweating is produced. Its diaphoretic property renders it valuable in colds, coughs, and in fact all cases where diaphoretics are needed. A valuable salve is made by bruising the root and stewing it in mutton or deer suet. The whole herb has a bitter, aromatic taste, not very disagreeable, and quite a pleasant smell.

SHELL-BARK HICKORY.

The Shell-bark Hickory is found in most parts of the United States, growing in strong, good soil.

The ross or outside bark of this tree makes one of the best diaphoretic teas or sweating medicines that we have. A decoction of the outside bark, not only acts as a sweating medicine, but is also good to correct the bile, and invigorate the stomach. Taken in very large doses it will operate as an emetic.

It is excellent to remove cold and female obstructions, and may be advantageously employed in any cases where a sweating medicine is needed.

PEPPERMINT.--(Munthea Piperita..)

Peppermint is a perennial plant, and is cultivated in most gardens; it is also found in many places in low, wet lands and is the strongest of all mints. It is said that the roots of the peppermint should be transplanted every three years, or it will degenerate into the flavor of spearmint.

Peppermint is a warm stimulant to the stomach, and through that medium to the rest of the body. A tea of it drank copiously promotes perspiration, and is useful to check vomiting, relieve hysterics, and remove sickness at the stomach. The essence of peppermint ranks among the best medicines in the world for expelling wind from the stomach; it is beneficial in allaying spasmodic affections of the stomach and bowels, dispelling flatulence or wind, and in removing all colicy pains. It will often relieve cramp, which sometimes takes place during the operation of an emetic. A few drops taken in spirits or water is very good to remove feeble feelings of a morning. Smelling it and wetting the temples with it will often relieve nervous head-ache. The green leaves infused in spirits or water, and applied to the pit of the stomach, and over the belly, will aid in checking vomiting, and in relieving spasmodic affections of the stomach and bowels.

SPEARMINT.

Spearmint grows in great abundance in most parts of America, on the banks of streams and in wet lands. It has a strong aromatic smell, and a warm, rough, bitter taste It possesses properties similar to those of the peppermint, but in a smaller degree. It may be used in decoction, oil or essence. The roots of the spearmint or peppermint, boiled to a strong tea and sweetened with loaf sugar is an excellent remedy for puking and purging in infants--(called cholera infantum.)

The spearmint is said by some to be an efficacious remedy for suppression of urine, gravelly affections, &c.; it is prepared by bruising the green herb and adding enough of the fourth proof Holland gin to make a saturated tincture--dose, a wine glassful taken as often as the patient can bear it. It is also said that cotton wet with this tincture and applied to the fundament, will give immediate relief in case of the piles. I have never tried it myself, but have mentioned it here, in order that those who wish to try the remedy can do so.

MOUNTAIN DITTANY.
[SUN-THEY-E-YOU-STEE.]

This useful herb is found in most parts of the United States, growing amongst rocks on hills and mountains. It has a yellow, fibrous, perennial root, and a smooth slender brittle stem, growing from six to twelve inches high, much branched at top, branches nearly opposite. Leaves are small, smooth and opposite, the upper surface of a deep green, and the under surface of a bluish green. Its flowers are numerous, small, of a pink white or a bluish purple, growing in terminal clusters.

The whole plant has a warm fragrant, aromatic, pungent smell and taste, and may be used in a warm infusion or tea advantageously in colds, head-aches, fevers, and in all cases in which perspiration is to be excited. Mountain Dittany is sudorific, tonic, stimulant and nervous, and may be used to advantage in all cases that require the use of such articles. It is very good for snake bite; in this case the tea should be drank freely, and the bruised leaves applied externally to the wound. A strong tea of this herb is valuable in increasing labor pains, and facilitating child-birth.

WINTER CLOVER, ONE BERRY, PARTRIDGE BERRY.

This is a small evergreen vine, lying close to the ground. It is mostly found in shady woods, often growing in beds or mats. Its leaves are small and round and grow out in pairs, the flowers grow out in pairs also, they are white and downy within, succeeded by light scarlet red berries. By the whites it is sometimes called Squaw vine from their having seen the indian women make much use of it.

The most common mode of taking it is in decoction, made by boiling it in new milk. It is diaphoretic, or produces sweating; as a diuretic it increases the discharge of urine; it is also slightly astringent and may be used to advantage in dysentery, piles, &c. The decoction taken freely is an excellent article to facilitate child-birth, it should be used daily for two or three weeks before that period.

BLACK DITTANY.
[TAH-TO-NE-GA-AH-TSE-HUS-SKEE.]

This herb is found mostly on dry stony ridges and hills, stem round, being much branched and growing from one to two and a half feet high, leaves are opposite, hairy and indented with unequal teeth so as to give it a very ragged appearance, the branches come out first above the leaves. It flowers in the latter part of summer and beginning of autumn, the flowers are tubular yellow on the out side, with a bright red spot within, the whole plant has a rough downy appearance, and when pressed by the hand it appears to be covered with a kind of rosin or sticky substance, it has but little smell or taste.

The leaves and branches are the parts used in decoction. it is a valuable diaphoretic, and may be used to advantage in all cases which require sweating medicines. It is one of the best articles in the world for child-bed fevers, when the stomach and bowels are prepared for diaphoretics or sweating medicines. It is good in colds, coughs, and most light family sickness. It increases the urine gently, also the menstrual discharges in females.

LONG-ROOT.

This root is perennial, large, branched--sending off finers. The root has pits or scars remaining on it where the old stem grew; stem round, smooth and shining, growing from two to three feet high--several from one root--generally found growing in rich uplands or bottoms.

The root is the part used; in a strong tea or decoction, it produces sweating--good in colds and obstructions. It is an excellent anodyne, and may be used to advantage in pleurisy and sharp darting pains; but in spirits it makes an excellent tonic bitter, useful in all cases where a bitter tonic is required. The tea sweetened is given to infants to relieve pain and produce sleep.

CEDAR.
[OH-TSE-NAH.]

This is a beautiful and well known tree, growing in most parts of the Union, particularly in the Middle Southern and Western States, and is too well known to require a description. It is a diaphoretic. A tea made of the leaves and twigs taken internally, at the same time stoving over the tops, is good for cold, female obstructions and measles. The berries boiled in sweet-milk will often expel worms. A very good ointment for itch and other cutaneous diseases, may be made by stewing the leaves and berries in sheep or deer's tallow, or hogs lard. But the most valuable medical property of this ornament of our forest has not yet been told. The oil of the tree is far the most useful part for medical purposes, and as an external application ranks amongst the best remedies for White Swelling, Rheumatism, and pains generally. It is also good to drive back bealings, and for diseases of the skin, such as ring-worm, tetter-worm, itch, scald-head, &c. For the manner of applying it in White Swelling and Rheumatism, look at the treatment of these diseases. The oil is obtained by filling a large pot with dry cedar finely split, and placing its mouth down wards on a rock; next exclude the air by means of mortar around the mouth of the pot, leaving a place at the lower side for the oil to run out at, then build a large fire on the pot. The oil may be used when fresh and after it has become old with equal success, if it is kept well corked. It should be kept in, glass bottles, or in well glazed earthen jugs, as it will readily penetrate wood. This oil put into the hollow of an aching tooth, will generally give relief in a short time.

PENNYWORTH.
[AH-TAH-NE-SEE-NE-TO-HA.]

This is a well known vine; it has a roundish leaf, inclined to be rough or hairy. It is diaphoretic. Good in colds and coughs: it is also good for children that are troubled with colic arising from cold. Either the green or dried leaves made into a tea is the form of using it.

GOLDEN ROD.
[TAH-O-NE-GA-AH-TSE-LUH-SKEE.]

Golden Rod is found in most parts of the U. S. with which I am acquainted, and is too well known to require a particular description. It grows from one and a half to two and a half feet high; they are long, narrow, of a deep green color, flowers numerous, small and yellow. The whole plant has a pleasant aromatic smell and taste. Taken in tea or infusion, it is good for colds, coughs, &c., owing to its diaphoretic or sweating property: it is also tonic, stimulant and nervine, valuable in measles, fevers, and female obstructions.

WATER BIG LEAF.
[OO-CAH-LO-KA-QUAH-AH-MY-QUAH.]

This plant is generally found in creeks, though sometimes in spring branches. The root is large, from two to four feet in length, with small or fibrous roots issuing from the main root: Generally two leaves grow up from one root; they always grow to the surface of the water, where they lie flat; they are large, roundish, fleshy, and of a dark green color. The root is the part used for medical purposes; it is diaphoretic, diuretic and antiseptic. A tea or decoction of this root, is a certain remedy for that dreadful disease Small Pox.

CLASS NO. VII.

DIURETICS.

DIURETICS are a class of medicines which are employed to produce an increased discharge of urine. They are valuable in all disorders of the urinary organs which prevent the secretion of the proper quantity of urine.

HORSE-MINT.
[SUN-STE.]

This plant is found growing almost everywhere, and is well known by almost everybody:--there are several kinds, all possessing the same medical properties.

The leaves and top are the parts used. A decoction of the Horse-Mint, midling strong, is excellent for weak bowels and stomach; it is diuretic, producing a free and easy discharge of urine in a short time after drinking it. As a diaphoretic, it is good to promote perspiration, which means sweating, giving relief in colds and female obstructions. It is also carminative, giving relief in cholic, flatulency and hysterics.

STRAW-BERRY.--(Frigaria.)

This is a small well known plant or vine, which bears a most delicious fruit of a cooling laxative nature. The fruit is the most useful part for medicine, but when it cannot be had, the vine may be used. It is valuable in diseases of the kidneys and bladder, and a good aperient in suppression of urine and viscernal obstructions; also in jaundice, scurvy, &c. The fruit, if held in the mouth for some time, is said to dissolve the tartareous concretions on the teeth.

TOBACCO.--(Nicotian Tobacum.)

This plant has been in use among the Indians or natives of the American continent time immemmorial, both as a luxury and medicine. It possesses at least eight grand medicinal properties. It is diuretic, emetic, cathartic, antispasmodic, sudorific, expectorant, anthelmintic, and errhine.

All these properties it possesses in a most powerful degree, yet its narcoctic effects on the system render it a dangerous medicine to tamper with. I have followed the example of some of my predecessors in classing this plant among diuretics, as it is generally given to act upon the urinary organs, notwithstanding it possesses some other of the above named properties in an equally active degree.

As a diuretic, it is not surpassed by any article in the compass of medicine. The urine discharged after the use of this plant, is entirely over the quantity of fluid taken into the stomach. This circumstance alone, is a conclusive proof that it acts powerfully on the urinary organs, and dislodges the dropsical fluid from the system: all the objection that can be urged against the use of this plant in the treatment of dropsy generally, is that of its being so active and powerful, as to require great caution and skill in administering it. I will give one remarkable cure of dropsy, which is recorded by different writers among the whites. It was performed by Dr. Cutbush, physician of the American Marine Hospital at Syracuse. The subject of this cure was a young woman, who had previously consulted thirty-three physicians of Italy, all of whom had given her over as incurable. The solemn entreaties of her parents determined her to make a trial, and as a last resort, he directed the leaves of Tobacco recently gathered, to be steeped in vinegar and applied over the abdomen. The first application produced sickness at the stomach, vomiting, swimming in the head, copious sweating, great depression of muscular strength, and a loose state of the bowels.

As soon as the above symptoms appeared, he removed the tobacco. This application he continued for several days twice a day, removing the tobacco, as soon as the above symptoms occurred, and in twenty days his patient was completely cured.

Where poisons have been taken into the stomach, which prevent the operation of emetics taken internally, tobacco leaves pounded and steeped in vinegar, or warm water, and applied over the pit of the stomach, will greatly assist the operation of the emetic taken internally.

ANTISPASMODIC.--In cramps, locked-jaw, spasms or colic, it seldom fails to give speedy relief, producing great relaxations of muscular powers, and unusual prostration of strength. Where colic is followed by obstinate constipation of the bowels, likely to terminate in inflammation, and consequent mortification, tobacco clysters may be used with the happiest effects, after the most powerful purges have been taken into the stomach, and have proved ineffectual--they should be given, one or two table-spoonfuls, in half a pint of new milk or thin gruel, repeating the clysters every half hour until relief is obtained, or sickness at the stomach produced.

As an ANTHELMINTIC, tobacco leaves applied over the stomach, have often removed worms when other remedies failed: it has even expelled the tape worm, but it is the most sickening application in the world, and should only be used as a last resort. This article may be used as an emetic by applying the leaves over the stomach in cases where laudanum has been swallowed for the purpose of destroying life. In such cases, the laudanum prevents the operation of emetics taken internally. As cathartic, it is used in clysters, as above directed.

SUDORIFIC AND EXPECTORANT.--It is never employed for these purposes, though it always produces these effects when exemplified for other purposes. The leaves cured in the common manner, is equally good as those recently taken from the stalk I believe. The tobacco steeped in vinegar, is the best application I have ever tried to the bee, wasp, and other poisonous insects, giving immediate ease to the pain.

RUSH. [CAH-NO-YAH.]

The Rush is an evergreen, growing in most parts of America. It grows in shady places along the banks of creeks and in the swamps of the South, and also in the cold and dreary prairies of the North west. The stalks are hollow, commonly about the size of a goose quill, and gradually taper from the root to the top, terminating almost in a sharp point. They have something like joints, and appear utterly destitute of leaves: it is said by some to be destitute of flowers and seeds also; as to the seeds I am not prepared to say, but I have seen it in bloom in the month of April. The branch or footstalk, which supports the bloom, puts forth from the joint towards the top of the stalk.

A decoction of the rush is a valuable diuretic, producing a copious and easy flow of urine, generally giving immediate relief in gravel. In the use of this article, the only directions for a dose is, be sure to drink enough, there being no danger whatever in its use. It is good for dropsy, taken daily in large quantities.

As a diaphoretic, it produces mild perspiration, or sweat. It may be gathered at any season of the year, and lay them where they will receive the fresh air, and they will keep sweet a great while: they are better for use when dry than when green.

SMART WEED. [OO-NE-TA-WE-TAG-TSE-KER.]   Of this weed there are two kinds, the great white and little red. The big white smart weed is perfectly inoffensive in smell and taste, accompanied with no strong sensation; but its stem and leaves are full of a slippery, mucilaginous substance. A decoction of this kind is diuretic, and is very useful in gravel and suppression of the urine, and very especially in strangury, or painful discharge of water. The decoction thickened with wheat bran or corn meal, forms an excellent poultice for swelled and inflamed parts, allaying the fever, and giving almost immediate relief.   The red or small kind, is very pungently acrid, and biting to the taste, and in appearance is about like the other, only a size smaller. Many persons place great confidence in this kind, as being a valuable remedy for discharges of bloody urine, but I have never tried it; it is useful in decoction or teas. The only use we make of it is a wash, and ointment for scald-head, as is fully shown under that head.

INDIAN HEMP. [CAH-TER-LAH-TAH.]   This plant is found growing in the woods, and on the borders of meadows. The stalk generally grows from 3 to four feet high, and is bare for some distance up; it then divides into several branches clothed with leaves; the Page 258 flowers are numerous, of a whitish color, similar to buckwheat, which are followed by the seeds, somewhat resembling a cucumber.   The root is the part used, and is a valuable article in the treatment of pox, when combined with other articles; for further information on this subject, refer to that head. An infusion of the root taken in gill doses, every three or four hours, operates as a diuretic, by increasing the discharge of urine, and is good in dropsy and uterine obstructions. It is also diaphoretic, or sweating, and will be found beneficial in rheumatism, asthma and coughs, especially whooping cough. If the dose be sufficiently increased, it will produce both puking and purging. As an emetic and cathartic, it is very severe, but in the latter way, it produces the happiest effects in pox when combined with other articles.

FLAX SEED.   This valuable article is cultivated extensively in many parts of the United States, for domestic purposes; the seeds are the parts used. Flax-seed is diuretic and expectorant. As a diuretic, it is good in gravel or burning in making water. As an expectorant, it is a valuable drink for persons afflicted with violent colds, coughs, and diseases of the lungs generally. A syrup made by adding a pint of honey to a quart of strong flax seed tea, and simmering it slowly over a gentle fire, for an hour, taking off the scum as it rises, is a valuable medicine for diseases of the breast and lungs, taken in doses of a table spoonful every hour, if the cough be troublesome.

SUMACH, BLACK AND WHITE.--Improperly called Shumake.--(Rhus Glambum.) [QUHH-LO-GUH, TAH-LO-HEE.] This is a shrub well known in the United States. Some practitioners prefer the white, others the black,--with us the white is generally used, though when it cannot be had the black is considered almost equal. The root of sumach forms a part of almost every preparation used by us for clap, and may be relied on as a valuble medicine in this dreadful complaint. The manner for preparing and using it in disease, is fully shown. As a diuretic, it acts well on urinary organs, and is valuable in strangury or painful discharges of the urine, and is also cathartic, antiseptic, tonic, and diaphoretic. The white shrub may be used as medicine: the leaves are good to smoke in case of asthma and phthisic. A decoction or infusion of the berries is a valuable tonic in ague and fever. The bark of the root acts on the bowels, as a purge, more actively than any part of the shrub. In cases of gleet or ulcerated bladder. A decoction of the root taken three times a day, a half pint at a time, is very excellent. The decoction forms an excellent wash for foul ulcers, a poultice made by thickening flour or meal in a strong decoction of the bark or roots, is a valuable application to risings; it either suppresses them or draws them to a head immediately, giving great ease to the pain. An ointment made by stewing the inside bark of the root of black sumach, in fresh butter, until the strength is extracted, is a most excellent application, for swelled or inflamed breasts.

SILK WEED.--(Asclepias Syriasa.) [OO-THUR-LOH-LAH.] The Silk-Weed, sometimes called Milk-Weed; is found in all parts of the country, growing mostly in rich grounds, the stem rises from two to four feet high, and bears a large pod, containing, when ripe, a silk like substance. The root of the Silk-Weed acts powerfully on the urinary organs when taken in decoction, and is valuable in venereal or clap. It acts well as a sweat, when combined with other articles, as is fully shown in the treatment of the different diseases. The root taken in large doses is emetic and cathartic; it is also tonic and is not surpassed by any root in my knowledge for a laxative bitter tonic. Persons afflicted with gravel or dropsy, will derive great benefit from its use; it has of itself without the aid of other remedies cured many cases of both dropsy and gravel. When gathered it should be kept carefully or it will loose its virtues.

WHITE-ELDER, SWEET-ELDER.--(Sambicus Niger.) [OOH-YOU-SUDK-UH.] The common Elder grows so plentifully and is so well known in this country, that any description whatever would be useless. The bark, flowers and berries all possesses medicinal virtues; they are diuretic or increases the discharge of urine, cathartic or purgative, emetic or puking. A tea or decoction of the inner bark has relieved obstinate cases of dropsy after other remedies had failed. The whites use this bark in tincture, made by digesting two handsful of the inner bark of the common Elder in a half gallon of wine twenty four hours. Dose, one gill twice a day, and increase the quantity if the stomach will bear it, until relief is obtained. We have never tried this tincture, but can bear ample testimony of the efficacy of the Elder bark in decoction. Digested in wine, it will be more palatable and grateful to the taste, and I presume equally as good as a diuretic or increaser of the urine. A decoction of the flowers, is a mild, sweating, purgative anodyne, very useful for light sickness among children. An ointment or salve, made by stewing the inside bark in lard or fresh butter, is a valuable application to burns and most eruptions of the skin--it may be rendered better for burns by adding to it while stewing an equal portion of the the root of bears-foot. PUMPKIN.--(Cucurbita Pepo.)   A decoction of the common Pumpkin-seeds is diuretic, or increases the quantity of urine, and is very good in gravel, dropsy and diseases which require medicines of this class. The oil of the Pumpkin-seeds is said to be much better than the decoction. I am not prepared to say from personal knowledge which is the best, but I copy the following from Dr. Smith, who professes to be well acquainted with its virtues as a medicine. "It is perhaps without exception, the most certain and most efficient diuretic we possess," giving immedite relief for the scalding of the urine and spasms of the urinary passage. Dose of the oil is from six to twelve drops, repeated as often as the violence of the symptoms require it.

QUEEN OF THE MEADOW.--(Spered Ulmaria.)   This plant is found growing mostly in wettish ground, though sometimes on high, dry land. It has a long fibrous root, which remains in the ground all winter, it is of a white or brownish color. In the spring several stems grow out from the same root from three to six feet high, they are round, smooth, jointed--of a purple color around each joint, bearing many pale reddish blossoms on the top in clusters. Its leaves grow out around the stalk at the joints in whorls, from three to five in a whorl, they are large and indented or jagged.   The root of this plant is the part used, and is a most powerful diuretic, useful in all dieases of the urinary organs, dropsy, gout, rheumatism and female obstructions. It is used in a strong decoction, and when taken freely is an almost certain remedy for gravelly complaints, as it seldom fails to carry off the calculus or stone with the urine in a dissolved state.

PARSELY.--(Apium Petroselium.)   This well known plant is cultivated in most of our gardens for culinary purposes, and therefore needs no description. The top and root are both used in form of a decoction; it is diuretic; the root is some the best. It is good in inflamation of the kidneys and bladder, in all ordinary cases of suppressed urine. A constant drink of this decoction has cured obstinate cases of dropsy at an advanced stage. It is one of the best teas in the world for infants afflicted with suppression or painful discharges of urine. It is also very good for female obstructions, and for lying in women, whose discharges are too scant.

CAT-TONGUE. [WE-SEK-KAH-CHAR.]   This plant has a clear smooth root, with but few fibres, running horizontal in the ground; about the size of a goose-quill, it is white, or of a clear watery transparent appearance, the root is not killed by the frosts of winter. Generally but one stem rises from each root, it is small, round, erect, covered with hair or down, and grows from one to two feet high. The leaves grow out opposite, alternate, Page 262 spear shaped, and like the stalk are covered with a down or hair--blossoms are white. The root is the part used, the best mode of taking it is in strong decoction. It is diuretic, or valuable in producing an easy and copious discharge of urine. In diseases of the kidneys and bladder, suppressed urine and gravelly complaints generally; it seldom fails in giving speedy relief. It is also good in dropsy.

TWIN LEAF.--(Jeffersonia Odorata.) [OO-NA-KER-OO-NAH-STA-TSE.]   Twin-Leaf has a small root, full of fibers, the root is perennial, that is, it is not killed by the frosts of winter. It has many leaves, which grow out on long foot-stalks, divided into two equal parts, which circumstance has given it the name of Twin-Leaf. The flower stalk produces a single flower, which is white. The most common mode of using this plant is in tea or decoction, but it may be employed in tincture or syrup if preferred. It is good for dropsy, suppression of urine, and gravelly complaints generally. It is an excellent external application, to sores, ulcers and inflamed parts.

WILD POTATOE.--(Convolvulus Panduratus.) [TE-CO-WE-SKEE.]   The Wild Potatoe has a large root, sometimes more than three inches in diameter and two or three feet long, branched at the bottom, of a rough appearance, having grooves running lengthwise. The root is of a yellow color, containing a milk like juice; it is perennial, that is the root remains in the ground all winter, and is not killed by the frost. Its stem is a climbing vine, running from three to twelve feet long, of a purplish color. Its leaves grow out alternate, and are somewhat fidddle-shaped, of a deep green on the upper and pale on the under side. The flowers resemble the morning-glory and are of a white or purplish color. This vine is mostly found in poor, loose, sandy soils, in open grounds.

The root is the part used, and may be taken in decoction or powder. As a diuretic, it is useful in dropsy, gravel and suppression of urine. It is a mild cathartic or laxative, Page 263 and expectorant, and is valuable in coughs, asthma or phthisic and consumption.

CLAP WEED, COCK-UP-HAT. [OO-STEE-CAH-NE-QUAH-LE-SKEE.]   The root of this plant is small, dark, and fibrous, and has an uncommon pungent, biting taste, and on being chewed produces a great flow of spittle. The stem grows from one to three feet high, leaves small, rough, pointed, rather oval. Its flower is about the size of a thumb, of a purple color and bears some resemblance to the sunflower, only it is more bulbous or round on the face than the sunfloower.   The root is the part used, and taken in decoction, tincture, or even the root chewed perseveringly, is a specific for veneral or clap in its worst forms. POOR ROBBIN'S PLANTAIN.   This plant is generally found growing in low grounds, it grows from 4 to six feet high, and often climbs on bushes near it. The upper part is white, armed with sharp spines or prickles. The flowers are small, succeeded by a fruit rather large, composed of two berries, slightly adhering together and covered with prickles.   The leaves are the part used, in decoction it is a most valuable remedy for suppressions of the urine and for all gravelly complaints. It is astringent and may be used for spitting of blood with the happiest effect, it is also good for epilepsy or fits. Dose, a half pint every hour until relief is obtained.

HIGHLAND BIG-LEAF. [OO-KAH-TO-GE-A-QUAH.]   This plant is found in great abundance in all parts of the United States with which I am acquainted. I do not know that it has any name by which the whites would know it. The Indians, when asked by the whites for a name, call it Highland Big-Leaf, but say they never saw it used by the whites for medicine. It has a large rough, woody root, which when broken, appears of a dirty purplish Page 264 or black color, it is very hard to dig; the root is perennial. The leaves are from three to five in number and put out from the root, supported on long foot-stalks, the stem or foot-stalk is generally of a purplish color. The leaf is broad at the base terminating in a point--the edges are indented with unequal scallops, the foot-stalks and leaves, particularly the under side of the leaves are covered with a kind of hair or down. During the summer it sends up a round naked flower stalk, from two to four feet high, which is crowned with numerous flowers, of a yellow color, succeeded by seed vessels not very disssmilar to those of the common tobacco stalk. It is mostly found on poor rocky lands and dry hill sides.   The root is the part used: it forms an ingredient in the antibillious pills, and is diuretic and cathartic. It is a valuable remedy for venereal or clap, as may be seen under that head. A decoction of it is good for gravel and diseases of the urinary organs generally.

HOT VINE.   This is a large, lengthy vine, turning itself to whatever is in its reach. It is a garden plant, and is too well known to require a further description. A strong infusion or tea of the hop is one of the most valuable remedies we have for gravel, and inflammation of the kidneys and bladder, it is narcotic or anodyne, that is it alleviates pain and produces sleep. It is good in rheumatism, and breast complaints, it gives tone and strength to the stomach, and invigorates the system generally. It is also valuable in female complaints, especially where the womb is debilitated, and for such females as are afflicted with the falling of the womb. It may be used in decoction or in spirits as the patient may prefer.

[OO-NA-STAH-LAH-CAH-TSEE-LE-SKEE.]   This plant grows in all parts of the United States with which I am acquainted. It grows on uplands and bottoms where the soil is good, mostly in fence corners. The root is hard to dig, has many roots issuing from one head, they are smooth, and inclined to be spindle shaped, with but few small fibres: on being exposed to the sun they turn of a Page 265 red color, and when dry, are wrinkled and dark. The bark of the root contains a kind of rosin or sticky substance, which in taste is something like pine rosin. The stem is erect growing from four to eight feet high, much branched towards the top, bearing numerous yellow blossoms. The stalk is round but has four and sometimes five welts, or feather like edges dividing it into so many equal parts, which gives it the appearance of a four or five square. I have never heard any name for it amoug the whites though it is growing in great abundance on many of their farms; they have never used it as a medicine, so far as I have been able to learn, unless it is the same weed that Turk recommends so highly for gravel, in a pamphlet published by him in 1843, he describes it as "having the appearance of a four square." The circumstance of its having sometimes five welts or feather edges might have been over looked by him, as every other particular mentioned by him is applicable to the Oo-na-stah-lah-cah-tsee-le-skee.   The root is the part used, and is one of the most powerful diuretics with which I am acquainted, it is an excellent remedy in gravel, and in all diseases of the kidney and bladder. It is also valuable for female weakness, such as flooding after or before child birth, in this case give it in tea or decoction until relief is obtained. This root should be kept at hand by every midwife, and by every physician who attends as a midwife, it is the most efficient remedy known to the Cherokees for floodings, and in their hands has never failed to give relief. For gravel or diseases of the urinary organs where there is not too much excitement, it may be taken in spirits as bitters, in this way it makes quite a pleasant drink. The quantity taken for gravel is not limited, only be sure to take enough. In female diseases, the quantity must be regulated by the effects produced.

 SOUTHERN YAUPON.   This is a well known shrub or bush in the southern states, it grows from twelve to fifteen feet high, its branches are numerous, upright and stiff, covered with a smooth whitish bark. Its leaves are evergreen, small and saw edged, flowers small, white, growing out irregular among the leaves, and are succeeded by small red berries, which become Page 266 red in the fall and remain so all the winter. The leaves of the Yaupon makes a pleasant tea, and when freely used, produces a copious and easy flow of urine and is a most valuable article in the cure of dropsy and gravelly complaints. The leaves toasted and made into decoction is a most valuable and pleasant drink. The Yaupon tea is as grateful to the taste as the Bohea tea, if not more so, and may be cured and preserved for use a great while, and is quite a convenient article of transportation.

BURDOCK.--(Arctium Lappa.)   This plant grows around rich yards, horselots, barns, and in other rich places, and is well known. The roots or seeds may be used. In decoction, it is a mild cathartic or purge; produces sweat, and a copious flow of urine. It is valuable to cleans the blood, and is admirably adapted to old venereal diseases, mercurial complaints, rheumatism, gravel, scurvy. &c. The root infused in spirits forms a valuable bitter for weakly females.

WILD RATS BANE, WINTER GREEN. [OO-NAH-TSA-LER-E-TRA-TA-KEE.]   This is a common plant throughout the Union, but is most abundant in the eastern and middle states, growing in sandy plains, and where pine timber abounds. Its root is woody, creeping, sending up stems at different distances. Its leaves are long narrow, wedged shaped, of a dark green color, variegated with light or whitish stripes, surface smooth and shining, with notched or indented edges. The flowers are white or of a light purple color, growing on the top of the stem, and are succeeded by brown seeds resembling allspice, it is an evergreen. The top and root are both used medicinally. In decoction or bitters it is diuretic, and will be found an excellent remedy for dropsy, and diseases of the urinary organs. A decoction of this plant is good for rheumatism scrofula, cancers, ulcers, &c. --it should be taken internally; in doses of half a gill, several times a day, and externally for bathing or washing the affected parts. Stewed in hog's lard it is said to cure tetter and ring-worm.

BEAR-BERRY, WILD CRANBERRY.--(Uva Ursi.) This is a low, evergreen shrub, which grows and spreads itself near the surface of the ground; its branches are pensile or hanging; bark is of a reddish or pink color, and is thickly set with oblong, oval, and entire fleshy leaves. The flower is oval shaped, broader near the base than the mouth, the edge of the flower is scolloped into five divisions, small, blunt and curled points. It produces fruit every two years; it is a roundish, red colored berry, similar in appearance to the small wild cherry, it contains five seeds and no more, they are hard, and with plain sides. This plant bears a strong resemblance to the bilberry or red myrtle, and good botanists often mistake the red-myrtle for the uva ursi. The only difference which can be depended upon, is this: the uva ursi has ten stamina or uprights, and the berries contain only five seeds each, while the bilberry or red myrtle has but eight stamina in the flower, and sometimes twenty seeds in the berry. Uva ursi is to be met with in the coldest countries and on the highest mountains, and elsewhere. The leaves and bark have a bitter astringent taste, and by those who have tried them are considered one of the best remedies now known for all diseases of the urinary organs, whether of the kidnies, ureters or bladder; many wonderful cures are on record, that have been performed by this wonderful plant. It is also good for dropsy--dose, a tea-spoonful of the powdered leaves in hot water, three or four times a day, or a decoction of the fresh leaves may be taken in teacupful doses several times a day. This article can always be had genuine in the shops, in this and most other countries where it is not to be found in the woods.

(ENGLISH NAME NOT KNOWN.) [AH-SQUAH-NA-TA-QUAH.] This valuable herb grows in rich mountains at the head of small streams--it grows from three to five feet high, the blades or leaves resemble the blades of corn; the roots many, small, of a dark, brown color.

The root is the part used for medical purposes. It is diuretic, and is an infallible remedy for dropsy. It also possesses the extraordinary property of reducing very fleshy persons down to their ordinary size, without impairing the health or effecting it in any way. The mode of using it, is the bruised root in cold water; it should be used daily, until the desired effect is produced: As it has no disagreeable sensation whatever, it will generally reduce the flesh sufficiently in the course of six months. The quantity taken each day is not at all material, but it should not fall short of a pint a day. The herb grows in great abundance on the Coahutta Mountain.

PINEY WEED. [NO-TSE-E-YAU-STEE.] This plant has a whitish fibrous root, stem erect, growing from five to eight inches high, and very much resembles young flax, and bears a purplish or whitish blossom on the top, found growing in pine and hickory soil. The whole plant may be used for medicine. It is diuretic and diaphoretic, the decoction taken internally and the bruised herb applied externally, will cure the bite of the copper head or rattle-snake; it is also good for venereal, conbined with highland big-leaf it seldom fails to cure that disgraceful disease. It may be taken in powders or decoction.

RATTLE-SNAKES' MASTER. [E-NAH-LE-UP-LOH-SCOCH-LA-NUR-WA-TEE.]

Rattle-Snakes master is found growing in considerable abundance in many parts of the United States. It has a perennial root nearly bulbous, about an inch in length, the lower end decayed giving off many fibres. Its leaves or blades put forth from the root, they are smooth fleshy, generally from five to eight inches in length, of a beautiful green color. The whole plant contains a kind of mucillage. The mucillage in the leaves will rope a considerable distance when the leaf is broken crossways. This herb is a most powerful stimulant a diaphoretic, it is also antiseptic. It is one of the most powerful and certain remedies for snake bite now known.

WILD MERCURY.

This herb is found mostly in cultivated lands, but sometimes in the woods. The main root is roundish, from 1 inch to an inch and a quarter in diameter, many small or fibrous roots issue from the main caudex or head, the whole root is of a brownish color--stem rises from two to four feet high, hairy and erect. Its leaves are large and pointed; flowers are yellow, with a round black middle--when the leaves of the flower fall off they leave a black burr. The root is the part used for medicine. It is diuretic, antisceptic and diaphoretic. Useful in dropsy, gravel and the like. This root seems peculiarly adapted to that disgraceful disease the pox, as may be seen in the treatment of that disease.

CLASS NO. VIII. EXPECTORANTS.

Expectorants are a class of medicines, used to promote the discharge of mucus or any other irritating matter from the lungs, and are useful in consumption, asthma, coughs, and in all cases where the excretions are not sufficiently active to throw off their contents. ELECAMPANE.-- (Inula Helenium.) This plant is found mostly along road sides, and about houses, it is also culvitated in some gardens; its leaves are large, flowers large and yellow, appearing in July and August. The root is the part employed medicinally, and when dry has an aromatic smell, and a bitterish aromatic taste, and not very unpleasant. It is a valuable stimulating expectorant, and is an excellent remedy in disorders of the lungs, as coughs, asthmas and consumptions. It promotes an increased flow of urine, acts gently on the bowels as an aperient. It is a strengthning restorative medicine, and a gentle diaphoretic. The root finely pulverized and mixed with honey is the best mode of using it--dose, a tea-spoonful morning and night; or it may be taken in syrup as follows: Boil a half pound of the root in a gallon of water, down to half that quantity, strain or filter that decoction, add a pint of honey or molasses--dose a table-spoonful taken morning, noon and night. It is also valuable for female obstructions, where the general health is impaired, and for pregnant women of weak habits, such as are afflicted with weak bowels and womb. It forms an ingredient in the chalybeate pill.

RED PUCCOON.--(Sanguinaria Canadensis.) [RED-LAH-NO-TA-HAH.] Sometimes called Indian plant, blood root, &c. This plant is a native of North America, from Canada to the Gulf of Mexico, and perhaps of no other portion of the globe. The root is perennial; it has a fleshy root, of a reddish brown outside, but on being broken, or cut, it emits a bright red juice, resembling blood, hence the name blood root; it is about the size of the little finger, very tender, and the lower end of the root has the appearance of being cut off or broken, in taking it from the ground. The leaves grow out from the root, and are supported on long foot stalks, heart-shaped, of a pale light green, streaked with veins of an orange color; it produces a single white flower, succeeded by numerous seeds, which are round and pointed. The leaves and seeds of the Puccoon plant, are poisonous, and must never be used. The root appears to contain all its medicinal qualities, and in many respects is closely allied in its effects on the human system, to the seneka snake root. A decoction of the root taken in small doses, acts as an expectorant, and is valuable in coughs and inflammation of the lungs; this decoction given in table spoonful doses every fifteen minutes, until it produces puking, is a most excellent remedy in croup, or the powdered root, may be administered in broken doses, until the desired effect is produced. This in powder from twenty to thirty grains, is an active emetic for grown persons. It is diaphoretic, that is, a valuable sweating medicine, when given in broken doses. The dried root in spirits, makes a good strengthening, or tonic bitter, and is much used in our practice; in this way when combined with other articles, such as rattle root, star root, &c.; in all cases Page 271 of obstructed menses; where the general health is impaired, and tonics or strengthening medicines are required. It is good in colds, pleurisies, rheumatism, liver complaints and other inflammatory disorders. The decoction is a good wash for indolent ulcers, and the dried powdered root sprinkled on ill conditioned sores, seldom fails to produce a healthy discharge, and a disposition to heal.--The root sliced and steeped in vinegar, eight or ten days, is a certain cure for tetter worm. It is said that the pulverized root, snuffed up the nose, will remove polypus, a fleshy or gristly substance, which grows in the nostril, gradually increasing in size until breathing becomes difficult, and sometimes unless removed, ends in suffocation. The best time for gathering the root for medicine, is when the seeds are ripe, which is in May or June.

LIQUORICE.--(Glycyrrhiza Glabra.)
This plant is a native of both Europe and America, and is said to grow spontaneously in the north western part of the United States. The root is the part for medicinal use; they are long and large, of a tough or woody nature, and have a little the appearance of spikenard roots; they are of a brownish yellow color, and when chewed, they yield a sort of waxy substance, of a pleasant sweetish, and at last, a slight bitter taste. Both the root and extract is useful in coughs, asthma and hoarseness, as it loosens the phlegm, and promotes expectoration. The extract is made by steeping the root in water, or by boiling it until the strength is extracted, then strain the decoction and reduce it to the proper consistence by boiling or simmering it over a slow fire.

(THE ENGLISH NAME NOT KNOWN.) [CULSAY-TSE-E-YOU-STEE.]
This plant is found growing in rich bottoms and along the borders of meadows. It has a fibrous, perennial root, of a pale or whitish color, stem rises from four to six inches high, is decumbent or bending, the leaves grow out on the sides of the stem, and when the stem is straightened or made stand erect, the leaves stand edgeways in the same manner as the tansy leaf; they are smooth, of a deep green color, somewhat pointed; at the extremity of the stem there are three leaves, the next two grow out opposite, and the remainder alternate. The whole plant and root is used, and in decoction is an infallible remedy for whooping-cough.

ONIONS. The Onion is extensively cultivated in the gardens of this country, as an esculent root. The onion is expectorant and diuretic, is valuable both as an internal and external remedy. The onion juice prepared by putting a small quantity of brimstone in the middle, or heart of the onion, then wrap in leaves or wet cloths and roast it perfectly, then press out the juice; this is excellent for hives and croup among children. The onion bruised and mixed with or stewed in lard, is one of the best external applications in the world for croup, colds, sore throat, phthisic, &c. The throat should be annointed and well rubbed with it in the above cases. When taken in any reasonable time, the juice taken internally, and the bruised onions mixed with or stewed in lard, applied externally, rarely fails in giving relief even in very obstinate cases, and this mode of treatment has proved successful where other remedies prescribed by physicians of high standing, appeared to have but little effect. Persons afflicted with liver complaint, will derive much benefit from the daily use of this root as food. It is said that a gill of the red onion juice taken morning and evening, a half hour after taking each dose drink a pint of horsemint tea, will entirely cure gravel and dropsy. The onion poultice applied to the feet in nervous fever, and acute diseases, is an excellent application to produce a revulsion from the head; it is also a good application to the throat and breast, in colds, where the throat and breast is much affected.

LINCHEN OR LUNGWORT.
This is a thin skin or shell, which grows on the bark of the white-oak tree, and is thought to bear some resemblance to the lungs, from which it has taken the name of Lung-Wort.

A strong decoction of this article made into a syrrup with honey, taken in gill doses, several times a day, is good for whooping-cough. It is also valuable in consumption and other diseases attended with a cough.

WHITE HOARHOUND.   This is a well known plant, the leaves are remarkably bitter and unpleasant to the taste. A syrup made by adding to a quart of the decoction, a pint of honey, and simmering it over a slow fire one hour, is good for hoarseness, colds, coughs, and breast complaints. HYSSOP.--(Hyssopus.)   This is a well known garden herb, and a tea of it is good in fevers and in most inflammatory cases for sweating; it is also good to bring on a regular flow of the menses. A syrup made of the strong decoction of expressed juice, mixed with honey, is a most pleasant and valuable expectorant, useful in all cases where expectorants are needed, as in colds, coughs, asthma, and other diseases of the breast and lungs, accompanied with inflammatory symptoms. In tea or decoction, it is a mild and pleasant drink. TAR.   The common Tar which is obtained from the rich pine roots or knots of the common pine tree of this country, is one of the most valuable expectorants in the world. The water of tar, is a valuable drink for persons afflicted with consumption or breast complaints. Tar is a principle ingredient for consumption, as may be seen in the treatment of that disease. The tar plaster is a valuable external application to scald-head, letter-worm, stone-bruise and full or ill-conditioned sores or ulcers, also for inflamed sores.

MAIDEN FERN, MAIDEN HAIR, SPLEEN WORT.   This weed grows mostly on stony, shady bluffs, to the height of from six to ten inches, stem slender, smooth, erect, of a dark black or purple color; the leaves grow in the same manner of those of the highland Fern, they are fine, soft, and of a pale green or yellowish color. Page 274

An infusion of this plant, made by pouring a quart of boiling water on the dry herb, sweetened with honey, is valuable for coughs and diseases of the breast generally; also for acrid humors, and irregularities of the menses.--Persons afflicted with the liver complaint will derive great benefit from its daily use. It may be prepared as above directed, and taken in tea-cupful doses several times a day, or combined with other articles.

LIVER WORT.   This little plant is so well known as to render a description unnecessary. It is expectorant and diaphoretic, it also acts on the liver. Persons laboring under a diseased liver will derive a great advantage from its use. The proper mode of using it is a tea of the leaves.

MILK-WEED, MILK-WORT. [OO-NA-KAH-OH-TSE-LUR-SKEEE.   This herb is found mostly in stubble lands. The root is about the thickness of a goose-quill or some thicker.--There are two kinds, white and black, they cannot be distinguished by the top, but by the root alone, they both possess the same medical proprieties. Its stem rises from two to four feet high, leaves small, of a beautiful green color, flowers white.   The root is the part used. In tea, decoction, or powders it is a good expectorant, it is also cathartic when taken in large doses; but it cannot be recommended for a purge being too severe and drastic. It forms an ingredient in the anti-billious pills.

CLASS No. IX. ANTISPASMODICS AND NERVINES.   Antispasmodics are a class of medicines employed to prevent or relive spasms of every kind. Nervines act on the same principle, though in a less powerful degree.

FIT-ROOT, ICE-PLANT.--(Monotropa Uniflora.)   This plant grows in the woods of the western country, generally to the height of six or seven inches. It is a most singular plant in its appearance, several stalks spring up from the same root and turn white in September; the stalk is like frozen jelly, and when held in the hand dissolves like ice.   The root is the part used; pulverized and given to children it has proved beneficial in curing fits. Either used alone or combined with antispasmodics it is useful in epilepsy and convulsions. Dose, half a tea-spoonful of the pulverized root every half hour until relief is obtained.--The expressed juice mingled with water, is a good wash for sores eyes.

GINSENG.--(Panax Quinquefolium.) [OH-TAH-LE-GAH-LE.]   The root is called by the people generally, San or Ginsang, though improperly. It is found in great plenty among the hills and mountains of Tennessee, mostly on the north side of rich shady hills and ridges. A few years since the root of this plant was exported to China, where it was sold for four times its weight in pure silver, and in 1784, the price at Pekin, is said to have been eight or nine times its weight in pure silver. The chinese attributed great virtues to this root--they considered it as a sovereign remedy in all diseases incidental to their climate and country, and had no confidence in any medicine that was not combined with it; thousands of lives were lost among the poor (as they superstitiously believed) for the want of it. The stalk is smooth, growing from twelve to eighteen inches high, divided at the top into three branches, each branch bearing five leaves. Its leaves are oblong, broadest towards the outer end and indented on the edge; its flowers are small and white, succeeded by a large red berry, root spindled-shaped, of a yellowish color. It is also good for weakly females and for weakness of the womb, and nervous affections, convulsions, palsy, vertigo and dysentery. In these last named cases, it may be taken in spirits if preferred.

MOCCASIN FLOWER.--(Cypripedium Luteum) [COOK-QUEH-OO-LAH-SU-LOW.]   The whites have a great many names for this plant' such as Valerian, Lady-Slipper, Umbil, &c. It is said to grow in every State in the Union, and in all kinds of soil. Several varieties of this plant are found, some smooth and some rough or hairy. From one to four or five stems put up from the same root, and grow from one to two and a half feet high, bearing each from three to seven leaves, which grow out alternate, sheathing the stem. The leaves have many nerves or fibres-running through them, giving them a rough uneven appearance. The flower in shape resembles a moccasin. The root is of a pale yellow or brownish color, having a large cluster of long, round, crooked fibres growing in a mat.   The roots are the only part of this plant used as medicine, and should be gathered in the fall, when the top begins to die, or in the spring, when it first puts up. It is a most excellent nervine and antispasmodic, useful in all cases of nervous irritation, spasms, fits hysterical affections, &c. It acts admirably well as an anodyne in relieving pains, quieting the nerves, &c. Its effects on the human system, in quieting the nerves, and relieving spasms, are superior to those produced by opium while it does not produce any of the narcoctic or stupifying effects of that drug. Dose a table-spoonful of the pulverized root, repeated as often as the case requires it--it may be combined with other medicines, and all act equally well in allaying and keeping down nervous irritation, or when used alone; this is one great advantage, for its treating disease, it is often necessary to give such medicines as produce nervous irritation, which difficulty may easily be overcome, by combining with such medicines a portion of the Mocasin flower root. When gathered, it should be well washed and dried, then pulverized, and kept in well corked bottles, or it will loose part of its virtues.

CHAMOMILE--(Anthemis Nobiles.)   The Chamomile is cultived in the gardens of this country, for family medicine, and is too well known to require a description. Its flowers have a strong, aromatic smell, though not unpleasant; they have a very bitter nauseous taste.   A strong decoction or tea made of the flower or herb is good in female obstructions, hysterical affections, colics, vomiting, bowel complaints attended with much pain, and nervous irritation; it is also a good stomachic or tonic bitter. It is excellent in poultices for obstinate and foul ulcers, hard swellings, &c.

BLUE-BERRY--(Caulophyllum Thalictroides.) [TE-SO-QUAH-LA-GAH.]   This plant has many names among the whites, such as blue-cohash, pappoose roots, &c. Its stem is upright, straight smooth and divided at the top in two or three branches; each bears three leaves, in the centre of which grow out its flowers, (their color not remembered,) they are succeded by dark blue-berries, which enclose a hard stone or seed. The root is yellow inside, brown outside, hard and irregular, having many small fibres. The root is the part used, and is perfectly safe, and harmless in its effects on the system, it may be employed in decoction, tincture or syrup, and is valuable in every species of fits, hysterics, collics and nervous irritation generally. It is a most excellent medicine for females in promoting child-birth, and allaying inflamation, and in preventing mortification of the womb. Pregnant women should use this root for two or three weeks previous to the expected time of child-birth, by this means that painful event will be rendered comparatively easy and safe. CATNIP--(Nepeta Cataria.) Catnip is a well known herb, and needs no description. A tea or decoction of this plant is good for female obstructions, hysterics, worms, spasms, colics and the like. It is a valunble external application in poultice to swellings.-- The bruised leaves steeped in vinegar, applied externally is good to ally inflammation, reduce painful swellings, &c. A syrup made by adding a pint of honey to a strong decoction of catnip is good for coughs and colds which affect the lungs. Page 278 ASAFOETIDA. The article commonly known in this country by the name of asafoetida is the resinous gum obtained from the roots of a plant, that is a native of, and grows spontaneously in the mountains of Persia, and is technically called Ferula Asafoetida. It has a strong, unpleasant smell, and a bitter, acrid, biting taste. That is reckoned the best which is of a clear or pale reddish color, variegated with a number of white pustules, like tears. Its action on the system is both quick and powerful, generally affecting speedy relief in spasms, cramps, flatulence and hysterical affections. In all spasmodic affections which are produced by a torpid or inactive state of the bowels it is a most excellent remedy. It sometimes relieves whooping-cough and asthma, its action on the system is similar to that of garlic, though much more active and powerful. It is stimulant and expectorant, and in some instances answers well as a worm medicine. Such as are troubled with frequent attacks of colic, may rid themselves of the return of this painful complaint by making daily use of this article in spirits, taken one or two drams a day, as bitters. Sucking infants whose mothers make a free use of this bitter, are apt to grow much faster and enjoy much better health, than the children whose mothers use no such bitters; such children are seldom troubled with colic or pains in the stomach, they have little or no use for laudanum, Bateman's drops, paregoric, and such like preparations, it may be given to children by dissolving it in water. It may be administered to grown persons in tincture, in pills, or dissolved in water; if taken in pills, one pill of the common size should be taken at a time, and repeated as often as circumstances require it; if the tincture is to be used, from a half to a whole tea-spoonful is the proper quantity, to be repeated at intervals of a half hour until relief is obtained. A sufficient quantity of this article put in whiskey to make it a proper strength for bitters is one of the best family bitters it the world.

CLASS NO. X. ANTISEPTICS. Antiseptics are a class of medicines that are used to prevent mortification from taking place, or remove it after it has taken place. The articles of this class, furnished by the vegetable kingdom, are both numerous and valuable, and are far more active and efficient, than any thing yet known in the mineral kingdom. It would be a very strange sight indeed, to see the red man of the forest, maimed with the loss of a limb to prevent mortification. Their knowledge of the active antiseptic properties of the herbs, roots, and plants of the forest, render such painful operations unnecessary. They have but to resort to the woods, gather, apply and be healed.

SLIPPERY ELM.--(Ulmus Albus.) [TAH-WAH-TSI-LUH.]   The white or slippery-elm is a well known tree of the American forest, and America produces no tree superior to it for medical purposes.   The inside bark is the part used. As a poultice nothing is superior, particularly to old sores, burns, and wounds, either fresh wounds or such as have become inflamed.-- The application of the elm poultice to fresh wounds or burns will, in almost all cases, be followed by a discharge of matter. When there is any appearance of mortification, the bark should be pounded, and boiled in water and made into a poultice, and applied cool. Taken internally it is cooling, and soothing to the stomach and bowels. It is an aperient or mild laxative, and may be used to great advantage by pregnant women that are troubled with heart-burn; in dysentery and chronic bowel complaints, there is nothing better than the slippery-elm jelly, made by pouring hot water on the inside bark. Where there are symptoms of inflammation in the stomach or bowels, this jelly should be used freely as injections and by the mouth, this mucilage or jelly is very nutritive, and will answer admirably as nourishment for those who have been borught very low by acute disorders. It is good for dyspepsy or indigestion, Page 280 quinsies, colds, catarrhs, coughs, consumptions. It is a certain and astonishing remedy in all bowel and breast complaints, and may be freely administered to children with the happiest effects.

LYNN TREE. [E-TAH-HAH.]   The Lynn is a native of America, and is found in almost every part, growing mostly in rich bottoms. Its wood is very soft, white and juicy. The inside bark and twiggs are the parts used: they afford a rich, well-tasted jelly, but little, if any, inferior to the slippery-elm bark. It is a most excellent article for pregnant women that are troubled with heart-burn, weak stomach and bowels. The jelly made by boiling the inside bark or twiggs, is good for coughs, consumptions, and in fact all cases where the elm is recommended, both for external and internal use. When the stomach has been over heated by the too free use of spirituous liquors, it is a most excellent remedy.

SHEEP SORREL, WILD SORREL. (Rumex Acetosella.)   This is a well known plant, growing in the woods and shady places, in every part of the country; its leaves and blossoms have a very pleasant, though extremely sour or acid taste.   The leaves and blossoms bruised and applied to old foul sores, have affected some wonderful cures, after many other applications had been tried to little or no purpose; it is very active and powerful in allaying inflammation, and producing a healthy discharge from the sore or wound, it is an excellent antiseptic and may be advantageously employed in all cases where articles of this class are needed.

GREEN PLANTAIN.--(Plantago Major.) [OO-STAK-KAH-YAH-TUH-TSE-SKEE.]
This plant grows in great abundance in most parts of this country, about yards, barns, and in fence corners.   The leaves bruised and applied to sores, inflamed wounds, bruises and the like, is good to reduce the inflammation Page 281 and produce a disposition to heal. The tea taken internally, and the bruised leaves or expressed juice applied externally, is an excellent remedy for poisonous bites and stings--if applied immediately after being bit by a snake, and the tea drank freely, it will in most cases give entire relief without the aid of further remedies. The tea or decoction is good for bowel complaints, and bloody urine, it should be drank freely, there being no danger whatever in it. The expressed juice is good for sore eyes.

BEAR'S FOOT. [CAH-TO-YAH-STEE.]   This plant is found in most parts of the United States, but is most abundant in the middle, southern and western states. The root is blackish without, and whitish within, resembling in size and shape, a small sweet potatoe; they grow several small roots from one main root or head, in a manner similar to the sweet potatoe, from which head, spring up several stems, or stalks, close together, growing from five to six feet high; leaves are about the size of a man's hand, but in shape they resemble a bear's foot, whence it has received the name of Bear's Foot.

A poultice made of the root of this plant, is very good to scatter bealings, and allay inflammations. The bruised root applied to burns, will extract the fire and give relief. It forms an ingredient in a very valuable salve, made as follows: Take of sheep's tallow, fresh butter, and a small quantity of beeswax, melt them together, to which add the bruised root of bear's foot, a small quantity of sweet gum rosin, and pine rosin, also the bruised root of heart leaves, stew them well together and strain for use. This salve will be found good for cuts, burns, &c. An oil made by stewing the bruised root of this plant in fresh butter or sweet oil, is good for ear-ache, it will give relief in some cases of pained joints, the ointment should be applied to the pained part, and bathed in with a warm iron. The root infused in spirits forms a valuable bitter for rheumatism. This root is said by some authors to possess the property of curing the white swelling. We have never tried it in white swelling, but we have cured cases of white swelling, where it had been perseveringly applied, and had utterly failed to do any good whatever. We Page 282 deem it a very good poultice for inflammations of the skin and flesh, but wholly useless in diseases of the bone, or the membrane which covers it. The bruised root stewed in hog's lard, gives relief in some cases of itch.

YELLOW-DOCK, NARROW DOCK. (Rumex Crispus.) [OO-NUH-TSA-THU-TSI-KEE.]   Dock is a well known plant growing about barn-yards, in fence corners, and about houses. There are two kinds, commonly distinguished as the wide and narrow Dock.--They possess similar medicinal virtues, the narrow is generally thought to be the best. The root taken internally in tea or bitters, is aperient, and an excellent corrector of the fluids. The bruised root made into a poultice, is a most excellent application to old sores or ulcers, hard tumors and inflamed parts. The root bruised and stewed in hog's lard, is useful in curing eruptions of the skin, itch, &c.--Drinking a tea of the bruised root while applying the above ointment, will greatly facilitate the cure. The roots and seeds made into decoction are good for dysentery, and bowel complaints generally. The expressed juice of the root applied twice a day for a length of time, will seldom fail to cure ring or tetter-worm.

SASSAFRAS--(Laurus Sassafras.) [CUN-STAH-TSE.]   This is a well known tree, in most parts of the Union. The root in a strong decoction, or infused in spirits, taken moderately, is excellent to purify the blood where acrimony of the blood exists--it is also good in diseases of the skin. It has often been sucessfully used in rheumatism; for this purpose, drink freely of the tea, and bathe the affected parts in tea of the same, until a high state of perspiration is produced, then wrap up warm in a blanket and cool off by degrees. The tea is a good drink in venereal. The root, bark and flowers, make a very pleasant and wholesome tea when taken in moderate quantities, but when drank to excess, it produces weakness and debility. A valuable ague pill is made by boiling the sassafras and mullen leaves to a strong decoction, then straining it, and Page 283 reducing it by slow boiling to the consistence of wax, or a proper thickness to make into pills. Dose: three or four pills morning and night.   A strong decoction of this root thickened with wheat bran or meal, and applied to inflamed wounds, or sores of any kind, will act powerfully in allaying the inflammation, and preventing mortification. The bark, or pith, taken from the stem or trunk, and steeped in cold water, is a cooling and a very pleasant wash for sore eyes.

MULLEN--(Verbascum.) [TSO-LAH-E-YOU-STEE-TUH-LO-NE-GA-AH.]   The mullen is a very common and well known plant, and therefore needs no description.
   The leaves boiled in sweet milk are a valuable remedy for bowel complaints, particularly among children. The leaves dried and pulverized, and applied to proud or fungous flesh, will destroy it. The leaves stewed in fresh butter, make a valuable ointment for the rectum, or gut, in case of the piles. The doction of the leaves is good to allay inflammation, and reduce swellings; for this purpose, bathe the affected part with the decoction, then make a poultice with the same and apply; this treatment should be followed until relief is obtained.

INDIAN BALM--(Trillium Latifolium.)   This plant has a short, thick, perennial root, resembling the Indian turnip; it is wrinkled, of a brown color, giving out many small fibres; the stem is smooth, erect, and grows from four to eight inches high. The leaves are oval, three in a whorl, growing at the top of the stem; it bears one flower on the stem above the leaves. The flower is bell-shaped, varying in color, according to the species, as red, white, purple; all possessing the same medical qualities; the flower is sweetened by a small berry that contains the seed. A decoction of the root of this plant is very good for profuse menstruation or flooding among females; it is useful in all kinds of hemorrhages, coughs, asthma, and bowel complaints. The pulverized root may be given in hot water, if prepared.   A poultice made of this root is good to prevent mortification, and it will be a valuable application to putrid ulcers, tumors, and inflamed parts.

BUCK TREE. [GEO-TLAH.] The beach is a native of North America and grows in great abundance in most parts of the United States. It is so generally and so well known, that the inhabitants of this continent would laugh to see a description of it. It is one of the greatest ornaments of the American forest, as well as one of the most powerful antiseptics known.--The principal use made of it, is in the form of poultices, made by thickening a strong decoction of the bark or leaves with wheat bran, rye or corn meal; it is good in all inflamed wounds, bealings, ulcers, &c., the part should be well bathed every twelve hours in the decoction, and the poultice made and applied as above directed. It is among the best applications that can be made to an inflamed or bealed breast.

POKE-WEED.--(Phytolucca Kecandria.) [TAO-YAH-TA.] The poke-weed is found in great abundance throughout the United States, growing in any kind of rich land. Its root is perennial; it is large, generally branching, covered with a very thin, brownish bark. Several stems spring up from the same root, growing very large, at first green, afterwards turning purple. It is one of the most grand, bold and beautiful looking herbs in America, when filled with ripe berries. The roots and berries have been employed with the best effects in old or chronic rheumatism. The sound ripe berries put into spirits with a little sulphur added, is a most valuable bitter for the rheumatism: it should be taken in drams three times a day. A poultice made of the roasted root is an excellent application to the feet in the last stages of nervous fever and acute disorders; it is also good, applied to ulcers, swellings, and rheumatic joints. A poultice made by thickening a strong decoction of the poke and buck-eye roots, with rye meal or wheat bran, is the best poultice for white swelling with which I am acquainted, as is fully shown under that head. The poke root is the principal ingredient in Turk's rheumatic ointment. The expressed juice dried in the sun to the consistence of a salve, forms a good plaster for cancers, and ulcerous ulcers.

BUCK-EYE. This is a well known tree in all parts of the U. States where I have been and thought to notice for it. The bark of the root boiled to a strong decoction and thickened to a poultice, is a very excellent application to inflamed parts, such as white swellings, sprains, tumors, &c. The only use we make of it is in poultices; taken internally it is poisonous, but as a poultice it is a most excellent article to extract the fever and prevent mortification.

BLACK SARSAPARILLA. [EE-LE-YER-SAY-HIS-TEE.] The black Sarsaparilla differs very much, both in appearance and medical properties from the white or yellow Sarsaparilla: the root is somewhat larger and of a darker color than the white or yellow Sarsaparilla. The vine is of an ash color, climbing bushes or saplings that may be in its way. Its leaves resemble the leaves of the apple tree: the root is the part for use, and should not be taken internally, as it might produce some unpleasant symptoms: the only use we make of it is in decoction, used as a bathe for the feet and legs, of persons addicted to swelling in those parts. It will allay the inflammation, and assuage the swelling in a short time, if properly attended to.

SPRUCE PINE. This species of pine has somewhat the appearance of the common Pine of the forest, and also of the cedar. The leaves grow in broad, flat fans or bats. Its seeds or berries are nearly like cedar berries to look at; they have a pungent, acid, bitterish, aromatic taste. A strong decoction made of the bark or leaves, and made into a syrup, is good for pregnant women that are troubled with a cough; it is also a good drink for catarrh or ulcer of the lungs. A strong decoction of the inner bark drank warm, is a very good sweating medicine in chronic rheumatism. The berries infused in spirits, is good for chronic rheumatism, and in venereal diseases. A fomentation of the Spruce Pine, is a good application to the swelled testacles, caused by the mumps, and also the swelled breasts of females, produced by the same cause. It is astringent, and stimulant or tonic. The oil, or essence is useful in colds, and externally in bathing preparations for pained joints.

SWAMP LILLY. [GE-GAH-GA-AH-TSE-LUR-SKEE.] The Swamp Lilly is a well known plant, growing in swamps and marshy places--sometimes about springs.--The root is the part used; bruised and boiled in water, and thickened in meal or wheat bran, is an excellent poultice for inflamed wounds, sores, bealings, &c.

HIGHLAND FERN. (Improperly called Highland Ferrin.) [YOH-NEX-TSA-STO.]
This plant grows in great abundance in many parts of the Union, and is generally well known. It has a dark perennial root, running horizontally in the ground, and when broken it is very mucillaginous. Its stem is smooth, green, growing from one to two feet high, dividing at the top into several branches, leaves green, growing out along the sides of the branches, similar to the tanzy leaves. The whole plant has a beautiful green and shiny appearance. It grows mostly on hill sides and on uplands; when the top is tender in the spring, it contains a mucilage similar to that of the root. It is a valuable application to inflamed swellings, wounds, &c. There are several modes of applying it:--the roots may be washed clean, bruised, and mixed with wheat bran and cold water, to the consistence of a poultice and applied, the face of the poultice being first smeared with sour cream: or it may be bruised, boiled and thickened with wheat bran and applied as above directed; or, the root may be well bruised, smeared with cream and applied. It makes a very cooling and pleasant poultice in all cases where there is inflammation. I use it a great deal in the treatment of white swelling, after it has commenced running, and still continues inflamed.

BALSAM OF FIR, OR SILVER FIR TREE OF AMERICA.--(Pinus Balsamea.)
This tree is very common in the northern climates; it is Page 287 also found in great plenty about the west end of North Carolina, and east end of Tennessee,

particularly on what are called the Smoky Mountains. It has somewhat the appearance of the white pine, and yields a most valuable balsam, which exudes from the tree like other turpentine, and is collected for medical use. When fresh it is almost transparent or clear, but after standing, it assumes a beautiful yellow appearance, and looks very much like sweet oil. It is known by the name of Canada Balsam, or Balsam of Fir. It is a valuable remedy for complaints of the breast and lungs, particularly where it is accompanied with pain soreness or cough. Dose: half a tea-spoonful at a time, twice a day. It is good for females afflicted with flour albus, or whites, falling of the womb, weak backs, &c.--Also for venereal, and diseases of the urinary organs. It is aperient, or loosens the bowels, and cleanses and heals internal ulcers. It is a very excellent external application to fresh wounds, I know of no article that will heal a fresh wound quicker than this balsam applied when first tied up. It is also a good application to ulcers, old sores, and the like; it forms an excellent ingredient in healing salves.

WILD INDIGO. The wild indigo is a large weed, resembling the common Indigo. A decoction of the leaves given in large doses is a good puke, in smaller doses it is a good purge. In poultices or fomentation, it is to allay inflammation and stop mortification. The root may be used in the same way

CHARCOAL, OF WOOD.--(Carbo Lingi.) Charcoal is a vegetable production, and is one of the most valuable and innocent medicines we possess. Prepared Charcoal is one of the most powerful antiseptics known in the compass of medicine. The best mode of preparing Charcoal is as follows: Take the common Charcoal well burnt of good sound green wood, (that used by smiths will answer as well as any,) reduce it to powder, put this into a vessel that can be tightly covered, raise the heat under the vessel until the coal becomes red hot, then take off the vessel and let it cool with the lid on. When the Coal Page 288 is cool enough, put it into bottles for use, the bottles must be kept tightly stopped or it will loose part of its virtues. It is a valuable article taken internally for dropsy and for costive habits. It is good in bowel complaints of a putrid nature. For internal use, great care should be taken to pulverize it fine, or it will irritate the tender surface of the bowels. It is an excellent antiseptic application to foul and ill-conditioned ulcers and mortifying wounds or sores.

VINEGAR. Vinegar is possessed of very strong antiseptic powers; as an external application, it is used to moisten antiseptic poultices, It is successfully employed to correct the putrid tendency of the fluids in putrid fevers, &c.

CLASS NO. XI. ANTHELMINTICS. Anthelmintics are a class of medicines used to destroy or expel worms.

THE CHINA TREE.--(Melia Azendarach.) The China tree is a native of China, and was brought from that country to America, and is now the common yard tree of South Carolina, Georgia, and many parts of Alabama and Tennessee. The bark, and especially that of the root, is one of the best worm medicines in the world. It should be boiled to a strong tea and sweetened to a syrup, and given in table-spoonful doses every hour untill the desired effect is produced. The fruit mashed and put into spirits and given to children every morning before eating is good to expel worms, the bark or bark of the root may be used the same way. The syrup acts well in removing worm fever. The pulp of the fruit stewed in lard is said to be go good for scald-head, ring and tetter-worm, and the like.

JERUSALEM OAK--(Chenopodium Anthelmintic.) [OO-SUR-GA-AH-SOO-YER-TI-MUH-WO.] This plant is said to grow in every State in the Union, and is too well known to require a description. Every part of this plant is used as a worm medicine, the roots, leaves and seeds are the best. The best mode of preparing it is by boiling the roots, leaves or seeds, in sweet milk, sweeten it with honey or sugar, a table-spoonful of this to be given before eating, morning, noon, night, and at bedtime give a large dose of castor oil or American senna to work it off; antibilious pills will answer quite well. The oil of this plant has been long considered a valuable medicine to expel worms, but the decoction in sweet milk is in my opinion equally good if not better.

CAROLINA PINK--(Spigelia Marilancici.) [GEE-GAR-GA-AH-TSE-LUR-SKEE.]
This plant is so well known to the people of Tennessee, and the Carolinians, as to render a description of it almost unnecessary. Its root is branched and very fibrous, its stem is smooth, erect, rising from one to two feet high, bearing long, smooth and oval leaves, outer points acute. The whole plant may be used, it is an acute vermifuge or worm medicine. The most common mode of using it, and as good a plan as any, is in decoction, about two ounces of the plant and root together when green, or about a half an ounce of the dried root, put into a quart of water, and boiled until tolerable strong, sweeten it well and to a child from six to eight years old, give from one to two table-spoosnful for a dose, repeated three times a day until relief is obtained.--Writers on this subject, appear to entertain various and contradictory opinions; some esteem it as a good medicine, while others pronounce it very dangerous indeed. The Cherokees have no idea how long this plant has been employed by them as a worm medicine, and that with the happiest effects. It possesses strong narcotic or stupifying properties, and when the system retains it too long, produces very alarming effects by stupifying the child, swelling the eyes and enlarging the pupils or sights of the eyes; for the above reason, it is better to administer it in Page 290 large than in small doses, it acts well on the bowels as a purge, and is not retained so long in the system as when given in small portions. It appears reasonable, that it must produce this sickening, narcotic effect on the worms; yet it does not expel or cause them to be discharged, and by retaining both the medicine and dead or sick worms in the bowels, even if the medicine should have no bad effects medicinally, all retained together would certainly excite fever, and produce evil consequences. But if given in large quantities it first acts on the worms, and secondly expels them by its cathartic powers. And even should you employ a large portion, and it should not purge, but act on the eyes of the patient in the above manner, you have only to discontinue the use of the pink-root, and give castor oil, to aid it in passing off. If the

fever should become high and the symptoms alarming, aid the operation of the oil by injections, and feel assured that the evacuations of the contents of the bowels will remove every unfavorable symptom, in a very short time.

WORM WOOD.   This plant is well known in this country, being generally cultivated in gardens for its medical virtues. The juice of the plant, sweetened with sugar, or honey, administered in table-spoonful doses, frequently repeated, and a poultice of the bruised plant externally on the stomach of the patient, is an excellent remedy for worms, especially where the bowels are much pained. It is also good for cramp colic. The decoction is good for grown persons troubled with hysterics and cramp colic, and for females afflicted with painful menstruation. A poultice made by bruising the herb or by boiling and thickening the decoction with wheat bran or corn meal, is good to prevent mortification, and heal up wounds, old sores, and the like.

COMMON GARDEN RUE--(Ruta Graveolens.)   This is a well known garden herb. The top or leaves boiled to a syrup with honey or sugar and given to children troubled with worms, is a most excellent remedy, it should be given every morning, in doses of from a tea to a Page 291 table-spoonful, according to the age and constitution of the child, when worms produce violent pains in the stomach and about the naval; a poultice of the bruised leaves applied externally over the pained part will greatly aid in giving relief. It quickens the circulation removes obstructions, promotes secretions. Persons troubled with hysterics will derive great benefit from the use of it in whiskey, as a bitter; it is also good in palsy where this disease is produced by debility or some obstruction.

Applied in poultices to the feet, in the last stages of acucte disorders it is excellent to produce a revulsion from the head, it will sometimes draw a blister. A poultice made by bruising the herb, or thickening the strong decoction with meal or wheat bran, and applied to inflamed or gangrenous parts, is excellent to prevent mortification.

DIRECTIONS, FOR SELECTING, GATHERING AND PRESERVING MEDICINES.

ANNUAL ROOTS.--That is, such as grow from the seed every year, should be gathered just before the flowers put out, as they are then in the highest state of perfection.

BIENNIAL ROOTS.--That is, such as grow from the seed the first year, live through the winter, and bear seed and die the second should be gathered either in the fall of the first year or spring of the second.

TRIENNIAL ROOTS.--Should be gathered in the fall after the leaves begin to die or in the spring before they put forth.   Roots intended for medical purposes are to be washed clean, and not kept long in the water as this will diminish their virtue; after being washed clean, spread them in the sun a short time to dry the water off, then spread them out in a dry place. When perfectly dry, pack them away to exclude the atmosphere.

Herbs and leaves intended for medicine should be gathered Page 292 about the time of flowering, as they are then at their greatest perfection. Flowers should be collected when in perfection. Herbs, leaves and flowers, must be cured in the shade.   Barks, designed for medicine, should be gathered either in the fall or spring, from young thrifty trees, shave off the rough or outside bark, dry it in the shade, and preserve it from rain and dew.

BLOOD-LETTING. It is highly necessary that the head of every family should understand how to open a vein with a lancet: they should also be acquainted with the cautions necessary for avoiding danger. Many cases occur where medical aid cannot be obtained in time, and where life is lost for want of bleeding. To draw blood from the arm you are to apply a bandage or ligature an inch or two above the elbow joint, and draw it so tight as to compress the veins immediately under the bandage, which will cause them to fill and swell immediately below it. As soon as the vein from which to draw blood rises, place the thumb of the left hand about an inch below the place you intend to open with the lancet; then with your right hand holding the lancet firmly between the thumb and forefinger, make the incision obliquely or slanting. This should be strictly observed, for by holding the handle of the lancet too high, the point will cut the under side of the vein and perhaps dangerously wound an artery. When the desired quantity of blood has been drawn, untie the bandage and place your thumb on the orifice and press it with a moderate force so as to bring the edges together. This will stop the blood from flowing. You are next to apply a compress about two inches square, made by 2 or 3 times folding a piece of linnen; over this you are to place a thick folding of linnen about four inches square, so as to fill up the bend or hollow of the arm. The folds of linnen is next to be confined by a tape or other bandage, which is to cross over them crosswise, extending above and below the elbow joint in the form of a figure eight, and to finish with making Page 293 a knot on the linnen immediately over the incision or orifice. If the bleeding should continue, pour cold water on the arm above the elbow; if this should fail to stop it take off the bandage, wash the orifice with strong vinegar and apply the cranes bill or some other styptic. To draw blood from the foot, place the bandage above the ankle joint and open the vein as directed for bleeding in the arm. The blood will flow more rapidly when the foot is emersed in warm water. Bleding, althho' indespensably necessary in many instances, may nevertheless be improperly resorted to. Regard must be always had to the strength, constitution and condition of the patient; stout, robust persons, of full habit, will require the loss of more blood than the more delicate and weakly. In all inflamatory diseases, bleeding will be found highly beneficial. But on the contrary where the disease arises from debility or weakness, bleeding will do serious injury.

DISPENSATORY. Having finished that part of Materia Medica, which embraces the simple articles of Medicine that have been introduced in this work in the treatment of the different diseases, and having pointed out their most obvious medical properties, together with the mode of preparing and administering them, I will now proceed to such compounds as the reader is referred to in the treatment of the different diseases.

This part of the work will be arranged in the form of a Dispensatory for the greater convenience of the reader.-- There is also given a short table of Mineral Medicines.

DIURETIC PILLS. Take one table-spoonful of copperas; one table-spoonful of pine rosin: the yolk of two hen-eggs: mix these articles well and add starch or flour enough to make it the proper consistence to roll into pills. Three is a common dose Page 294 for an adult. These pills are valuable for dropsy and other diseases that require the use of diuretics.

ANOTHER VALUABLE PREPARATION. Take of egg-shells half an ounce; coperas one ounce; elder-bark, of the root, one ounce; white sumac, bark of root, one and a half ounces; pine rosin one ounce. Brown the egg-shells and pulverize all the ingredients, and add water sufficient to the proper consistency to roll into pills.--Three or four is the common dose for grown persons. Used for the same purposes as the foregoing.

DIURETIC POWDERS. Take of the common elder, bark of the root, six ounces; burdoc root six ounces; egg shells browned, four ounces: queen of the meadow, six ounces; agrimony, six ounces: horse radish four ounces. Reduce all these articles to fine powder, and sift them through a fine sieve; then bottle up for use; dose, a teaspoonful three or four times a day, in a tea of water-melon or pumpkin seeds. These powders are useful to relieve suppressions of urine, and also to carry of the dropsical fluid from the body. When all the above articles cannot be had, use such as can be obtained; by turning to the materia medica, in the class of diuretics, you will find a number of simple articles described, any of which may be used in forming a compound, for gravel or dropsical affections; they should be pulverized and used as directed for the above preparation. Diuretic Tincture.--Take of elder leaves, one ounce; horse radish one ounce and a half; welt weed, the root, one ounce. Digest the whole four or five days in good spirits, in sun heat, shaking it well every day; of this tincture, take three or four drams a day; excellent for gravel, and urinary obstructions. Diaphoretics.-- Medicines of this class are employed to promote prespiration Diaphoretic Powders.--Take of butterfly root, one lb.; silk weed root, one pound; rag weed root, one lb.; seneka snake root, five ozs.; ginger one lb.; cloves four ozs.; red pepper two ozs.; reduce all these articles to fine powder and sift through a fine sieve, then mix them thoroughly. Dose for an adult is one tea-spoonful in hot water. This compound is useful in colds, obstructions and in the first Page 295 stage of disease generally. They give tone and strength to the system, promote prespiration and thereby discharge the morbid matter, and remove obstructions. Where the the above articles cannot be conveniently obtained, the reader has but to turn to the class of diaphoretics in Materia Medica, where he will find a variety of articles of this class described at length, any of which may be prepared and used as the above in all cases where the above would be of service. In preparing compounds as above directed, the ginger, cloves and cayenne should never be omitted: these are articles which can generally be had anywhere almost.

DIAPHORETIC DROPS OR TINCTURE. Take of blue-flag root, green, one ounce; cayenne one ounce; peach kernels one ounce; common ginger, one and a half pounds; gum-myrrh one pound; alcohol six quarts. Pulverize all the solid ingredients, and add them to the alcohol; digest eight or ten days in sun heat, shaking several times a day. Dose from one to four tea-spoonfuls, repeated at discretion; it should be taken in some kind of sweating tea. It will be found valuable for pains in the stomach, dysentary, colic, colds, head-ache, internal inflamation, &c.

CATHARTIC PILL. For the simple articles of this class, see "Cathartics," Materia Medica, where you will find them described. Antibillious Pills.--Take of gulver root 1 1-4 pounds; Indian physic 1 1-2 pounds; milk-weed root 1 1-4 pounds; highland big leaf 3-4 of a pound; black ash bark 3-4 of a pound; black walnut bark 1-2 pound; white walnut bark 15 pounds. Boil all these ingredients until the strength is extracted, then strain the decoction, and continue boiling until it is reduced to the consistence of very thick molasses; when it is cooled it will be stiff, add starch enough to roll it into pills. Eight pills of common size is a dose, if they fail to operate in a reasonable time, give half a dose. Gulver Pill.--Take any desirable quantity of gulver root, reduced to fine powder, add enough of a strong decoction of the same to make it the proper consistence to make into pills--from five to eight for a dose, repeated Page 296 in four hours, if they should fail to operate in that length of time. These pills are peculiarly adapted to nervous complaints. Gulver Syrup.--Boil any desirable quantity of this root to a strong decoction, strain and continue boiling until it is very strong, then sweeten it with molasses, and administer in table-spoonful doses, repeated every three hours until the desired effect is produced. Butternut Syrup.--Take any desirable quantity of white walnut bark, to every fifteen pounds of this bark add 1-2 pound of gulver root and a half pound of Indian physic, boil all the ingredients until the strength is extracted, strain and continue boiling to the consistence of molasses, then cool and bottle for use. Dose from one to two table-spoonsful. Hepatic Pill.--Take any desirable quantity of boneset leaves, boil them until the strength is extracted, then strain the decoction and continue boiling until it becomes thick, taking care not to scorch it, then add starch enough to enable you to roll it into pills. Three of these pills is a common dose. Useful for complain's of the liver. Antidyspeptic or Hepatic Syrup.--Take any desirable quantity of bone set leaves, boil to a very strong decoction, strain and boil to the consistence of molasses, add to each pint of this extract, a half gill of sourwood molasses. Dose: from a tea to a table-spoonful, morning, noon and night. Useful in complaints of the liver, indigestion, &c.

EMETICS. Emetics are a class of medicines used to produce vomiting; their operation may be rendered more easy and efficacious by the use of warm water after the first motion to vomit. Emetic Decoction.--Take of Indian physic or American ipecacuanha one pound, gulver root a half pound, put into one gallon of water, and boil down to a pint, strain and bottle for use; of this, give a half gill every twenty minutes until vomiting is produced. This class is among the mildest, safest and most certain emetics. These articles may be reduced to powders, combined in the same proportions. Dose of the powders, two tea-spoonsful to a gill of boiling water, repeated every fifteen minutes until vomiting is produced. Tincture of Lobelia.--Take of Lobelia, if dry four ounces, if green, eight ounces; alcohol one quart, pulverize the herb, and add the alcohol, digest eight or ten days in sun heat, shaking it several times a day. Dose from a tea to a table-spoonful, repeated every fifteen minutes, until vomiting is produced. The tincture made in the same manner of the pulverized seeds, is the strongest preparation of this herb. The tincture of lobelia is the safest and best emetic known for snake-bites and other poisons. Lobelia may be administered in powders. Dose a tea-spoonful every ten or fifteen minutes until vomiting is produced; they should be taken in some sweating tea, not over blood heat, as anything over this temperature will destroy the virtue of the lobelia.

BITTER TONICS.   Articles of this class are generally employed to assist the organs in recovering a healthy, vigorous action. The proper time to administer medicines of this class is after the force of disease is overcome by other remedies. The simple articles of this class are very numerous. I will here mention some of the best and most common:-- Bitter-root, hoar-hound, black and sampson snake-root, spikenard, columbo-root, goldenseal, bearberry, poplar bark, gum myrrh. These articles may be found under their proper heads in Materia Medica, where they are spoken of at greater length, their virtues may generally be increased by combining with them a portion of some astringent tonic. The following are some of the simple articles of this class: Dewberry and blackberry root, dog-wood, sumac, witch hazel, wild cherry tree bark, black birch, hayberry, cinnamon. When the use of astringent tonics produce dryness of the mouth, they should be discontinued.   Tonic Powders.--Take of columbo root six ounces, poplar bark, of the root six ounces, dogwood bark, of the root six ounces, wild cherry tree bark six ounces, boneset leaves four ounces, cayenne pepper one ounce, pulverize and sift through a fine seive, and mix all the ingredients well together. Dose, a tea-spoonful in either warm or cold water, repeated several times a day. Useful in debility, and in all cases where bitter tonics are required, as in intermittent fever or ague and fever. Where the above ingredients are not at hand, by turning to "Tonic" in Materia Page 298 Medica, you will doubtless find the description of such as are at hand, and such perhaps as will answer equally well. Where there is no fever the simple articles may be digested in spirits and taken as bitters.   Laxative Bitter Tonics.--Compounds of this kind should be formed of such tonics as possess aperient properties, by examing Materia Medica you will see several valuable articles of this class described, among which are poplar bark, boneset, sarsaparilla, black root, &c. They should be prepared and used as directed for tonic powders.   Chalybeate Pill.--Take of pleurisy root a half ounce, spikenard root a half ounce, star root a half ounce, elecampane a half ounce, Jerusalem oak seeds a quarter of an ounce, seneka snake root a half ounce, flour of sulpher three-fourths of an ounce, steel dust one ounce, pulverize all the ingredients, and sift through a fine sieve or thin cloth, then add a sufficiency of honey to cement it, and roll it into pills. Valuable in all cases where tonics are necessary.   Ague Pill.--Take equal quantities of mullen leaves and red sassafras bark, of the root, boil in water to a strong decoction, then strain and continue boiling the decoction to the consistency of very thick molasses, add a sufficient quantity of sassafras bark finely pulverized to make it the proper consistence to roll it into pills. Dose three or more morning and night. Useful in ague and fever.

NERVINES   Are medicines employed to allay nervous irritation, and should be employed in all compounds which are to be used in cases of this kind. One of the best nervines in Botanical Materia Medica is the root of the moccasin flower in doses of a tea-spoonful in warm water, repeated at discretion.   Nervine Powders.--Take of yellow moccasin flower root six ounces, ginseng one ounce, agrimony one ounce, nutmeg I ounce; pulverize all these ingredients, and mix them well together. Dose, a tea-spoonful in warm water, repeated as often as necessary. For a full description of several articles of this class, turn to "Antispasmodics" in Materia Medica, any of the articles in this class may be prepared and used as the one above, or they may be tinctured if desired, this is done by pulverizing as above directed, and then digesting in alcohol several days, shaking it every day.   Nervine Tincture.--Take of moccasin flower root six ounces, ginseng root two ounces, mountain dittany two ounces, sassafras, bark of the root four ounces, nutmeg one ounce, gum camphor a fourth of an ounce, alcohol three pints; pulverize all the solid articles, add the alcohol, and digest six or eight days in sun heat, shaking it every day. Dose, from one to three-tea-spoonsful, repeated every fifteen minutes until relief is obtained. Useful in colics, pains in the stomach, and very valuable for children.   Antispasmodic Tincture.--Take of moccasin flower root two ounces, cayenne pepper two ounces, blueberry root two ounces, asafoetida a half ounce, lobelia seeds a half ounce; pulverize all these articles, then add them to 1 quart of good spirits of any kind, let them digest eight or ten days in sun heat, shaking it well every day. Dose from a half to a teaspoonful, repeated as often as circumstances require. Useful in fits, spasms and the like: also in snakebites and where poisons have been taken into the stomach. Any of the articles in Materia Medica in the class of antispasmodics may be tinctured and used as above directed, but may be rendered better by adding a liberal portion of red pepper, and a small portion of lobelia seeds. Any of the articles may be used alone if others cannot be had, or the articles may be tinctured separately and the tinctures mixed.

ANTI-EMETICS.   Anti emetics are medicines employed to allay irritation of the stomach and check vomiting. When spontaneous vomiting proceeds from a foul stomach, means should not be used to check it, until the stomach has been cleansed with an emetic. The best articles of this class are cholera morbus root, seven bark, pepper mint and spear mint; there are also other articles which possess the property of anti-emetics in a less active degree; by examining materia medica, you will find a full description of the above articles, together with several others. They may be taken in tea or infusion, or tinctured in French brandy. The oil or essence of pepper mint may be used in tea for the purpose of checking vomiting. Medicines employed for this purpose, should be administered in small doses, repeated at short intervals, until the desired effect is produced.--Where vomiting is attended with spasms in the stomach, some antispasmodic should be combined with the antiemetic. The anti-spasmodic tincture will probably answer best for this purpose.

SEVERAL VALUABLE PREPARATIONS FOR VENEREAL. Make of highland big leaf root three pounds, piney root three ounces; put both articles into four gallons of water; boil down to one gallon, strain and bottle for use. Dose, one gill, three times a day. Another preparation.--Take of the white sumac root, one pound, dew-berry brier root, one pound; pine, inside bark root, one half pound; boil all these articles in 4 gallons of water down to two gallons; of this, the patient should drink a pint each day. Take white sumach root, two pounds; may apple root, one pound; devil shoe string, one half pound; persimmon, bark of the root, one fourth pound.--Boil in four gallons of water down to two; strain, and bottle for use. Dose, a half gill three times a day. Dr. Wright's beer for consumption.--Take of Spikenard root, if green, two pounds, if dry, one pound; seneka snake root, two ounces; wild cherry tree bark, one half pound iron weed root, one half pound; wild potatoe, one half pound; burdock root, one half pound; boil all these articles into ten gallons of water, and boil down to three: while boiling, pour the decoction into a keg or jug, and add one quart of honey; let it remain until it ferments, and it is fit for use. Dose, a half pint two or three times a day It is valuable in liver complaints, consumption, &c.

RHEUMATIC OINTMENT. Take of Cedar oil, one half pint, British oil one half pint mix well and anoint the affected part twice a day bathing it with a warm iron.
Relaxing ointment.--Take of Turkey buzzard's oil, one gill; fox oil, one gill: cedar oil, one gill; mix all well together. This is a valuable ointment for stiff joints; it should be applied to the affected part two or three times a day, bathing it with warm iron.
Essence of Pepper.--Take of African cayenne one-fourth Page 301 ounce, alcohol one quart, add together and burn one third of the alcohol away, then strain and bottle for use. This is a valuable external application to pained parts; it seldom fails to give speedy relief. Black Poultice.--Take of common soap, one fourth pound, hog's lard one fourth pound, table salt, three ounces, extract of sour wood, one gill; cedar's oil, one fourth gill; mix all these ingredients together, and apply in the form of a poultice. This poultice forms a good application to swellings, boils, stone bruises, and the like, drawing them to a head, and causing them to break, and run, much sooner than they would otherwise do. The sourwood extract, and cedar oil have a tendency to mitigate the pain.
Ointment for sores.--Take of bear's foot, the root, half pound, heart leaf the root half pound, elder bark, fourth of a pound; boil all together, until the strength is extracted, then strain and continue boiling to a very strong; then add to it of fresh butter, hog's lard, or mutton or deer's suet, two pounds; pine rosin, two ounces; sweet gum rosin, two ounces; stew all together, until the water is evaporated; then cool, and it will be ready for use. Valuable for cuts, wounds, sores, &c. Another for the same.--Take of tag elder bark, one half pound; common elder bark, one half pound, bamboo brier root one half pound; heart leaf root, one half pound; boil until the strength is extracted; then strain, and add to the decoction, of bees-wax, one fourth of a pound; hog's lard, one

pound; pine rosin, two ounces; sweet gum rosin, two ounces; stew the whole together, until the water is evaporated. To make a very drawing salve, add to either of the above, a portion of blue flag root, or balm of gilead buds.   Healing Salve.--Take of bear's foot root, a half pound; heart leaf root, a half pound; bruise and boil until the strength is extracted, then strain and add a half pound of fresh butter; stew it until the water is evaporated. Useful for eruptions of the skin, and in all cases where a cooling, healing salve is required, when both the articles cannot be had, either will answer.   Turner's Cerate.--This ointment, which is so celebrated in burns is prepared as follows: Take of calamine in fine powder, one half pound; bees wax one half pound; hog's lard, one pound; melt the wax with the lard and put it out Page 302 in the air, when it begins to thicken, or become cool, mix with it, the calomel, and stir it well until cold.   When you inquire for this article at an apothecary or Doctor's shop, ask for calamine in powder; it is a mineral imported from different countries. PLASTER FOR DRAWING

BLISTERS.   Take the root of wild wet-fire, bruise and wet it with water, and apply where the blister is desired; the skin should be moistened with vinegar also.
    Another for the same.--Take the bruised herb of wild camomile and apply it to the skin, as above directed.   Another for the same.--Take mustard seed, pound or grind them fine and make them into a plaster by wetting them with vinegar or spirits. Moisten the skin with vinegar or spirits before applying the plaster. Apply to the feet and wrists in the low stages of disease, to raise the pulse, and produce a revulsion from the head.
    Strengthening Plaster.--Take of pine rosin, obtained by boiling a rich pine root, 1-2 pound; African cayenne, 1-4 ounce; moccasin flower root, 1-2 ounce. Pulverize the cayenne and moccasin root, and add them to the rosin while warm, and mix them well, then spread it on a piece of thin leather or stiff cloth and apply it while sufficiently warm to adhere to the skin. When a sufficient quantity cannot be obtained by boiling the roots, that which exudes from the tree may be added, it soft. Useful for weak backs, also good to remove pains in the side, breast and back. It should be applied immediately over the pained part, and let it remain until it comes off.

STYPTICS.   STYPTICS are articles applied to wounds, cuts, &c., to stop the flow of blood.   Cranes Bill is a most powerful Styptic.--A full description of this herb, together with the mode of preparing and applying, may be seen under that head. It is said by some to answer equally well when pulverized and applied to the bleeding surface.
    White Hickory.--The inner bark boiled until the strength is extracted, then strain and continue boiling to the Page 303 consistence of molasses--forms an excellent application to stop bleeding; it should be applied by wetting lint in the extract, and applying it to the bleeding surface. It may be preserved any length of time, by adding to it a portion of good rum, and excluding the atmosphere by stopping it up in glass bottles. It forms a very good dressing for wounds, where there is not too much inflammation.
    Burnt Stone, finely pulverized and applied to a fresh cut, will in most instances stop bleeding. Persimmon inner bark, boiled to a very strong ooze, is very good to stop bleeding. The extract of Oak, either kind, applied to fresh wounds, or bleeding surface,

is good to stop the flow of blood. Soot applied to a fresh wound, is very valuable to stop bleeding. Sassafras Leaves bruised or pounded fine, is good to stop bleeding. In the class of "astringents" in Materia Medica, the reader will find several valuable articles for this purpose.

GLOSSARY, OR EXPLANATION OF THE TECHNICAL TERMS. Abdomen, lower part of the belly. Abortion, expulsion of the foetus before the 7th month. Abcess, a tumor containing matter. Absorbents, 1st. medicines that correct acidity, and dry up superfluous moisture: 2d. small delicate vessels that absorb fluid substances, and convey them to the blood. Absorption, the act of sucking up substances. Accoucher, one who assists at child-birth, a mid-wife. Acid, that which imparts a sharp or sour sensation. Acrid, burning, pungent, corrosive. Acute, a term applied to a disease denoting violent symptoms, hastening to a crisis. Adult, a person full grown. After-birth, the fleshy substance that connects the foetus to the womb. Affusion, pouring one thing on another. Ague-cake, enlargement of the spleen. Alimentary canal, the stomach and intestines. Alcohol, rectified spirits of wine. Alkali, any substance uniting with an acid neutralizes or destroys its acidity. Alternate, changed by turns; in botany, leaves and branches are said to be alternate, when they grow out singly on opposite sides of the stem, rising above each other in regular order. Amputation, the act of cutting off a limb. Anatomy, the dissection of organized bodies. Annual, yearly, every year. Anodyne, medicines which ease pain. Anti-acid, that which destroys acidity. Anthelmintics, medicines which remove or correct the bile. Antidote, a medicine that destroys poisons. Anti-emetic, a remedy for vomiting. Anti-scorbuctic, preventing or curing scurvy. Antiseptic, that which prevents or removes putrefaction. Antispasmodic, remedies for spasms. Amus, the fundament. Aperient, opening. Aarta, the great artery of the body. Artery, the canal conveying the blood from the heart to all parts of the body. Aromatic, fragrant, spicy, pungent. Astringent, medicines to correct looseness and debility, by rendering the solids denser and firmer. Axillary, in botany it means the angle formed by a branch with the stem, or by a leaf with the stem or branch.

Biennial, a botanical term applied to those plants which produce their roots and leaves the first year, and produce their fruit the second, and then die. Bile, the bitter, Page 305 yellowish fluid secreted by the liver. Bitternate, having three. Botany, that part of natural history which relates to the vegetable kingdom. Bulbous, a botanical term, denoting a round, oblate shape like that of an onion

Calculi, small stones or gravel. Caloric, the chemical term for the matter of heat. Calyx, a cup, the external covering of an unexpanding flower. Cancer, small corroding ulcers. Capsule, the part of the plant containing the seed. Carminative, that which expels wind from the stomach. Cataplasm, a poultice, soft plaster. Cartilage, a white elastic substance connecting the bones.--Catarrh, a discharge from the glands about the head and neck. Cathartic, a purgative medicine. Catheter, a small tube for drawing off the urine, being introduced into the bladder. Caudex, a botanical term denoting the main head or body of a root. Caustics, burning applications. Cellular, consisting of cells. Chancre, a venereal ulcer. Chronic, a term denoting a disease of long standing. Clyster, a liquid substance injected into the bowels. Connate, growing from one base, united together. Coagula, clots of blood. Conoption, the impregnation of the womb. Constipation, great costiveness. Constriction, a drawing together, contraction. Contagious, caught by infection. Cordate, having the shape of a heart. Corrosive, consuming, eating away. Convalescence, the state of returning health after sickness. Convulsion, a violent spasmodic affection, a fit. Corymb, a cluster of flowers at the top of a plant forming an even expanded surface Cutaneous, belonging to the skin. Cyme or cyma, an aggregate, like the sunflower.

Decoction, a preparation by boiling. Decumbent, reclined, bending down. Delirium, craziness, alienation of mind. Detergent, cleansing. Diaphoretic, promoting sweating. Digestion, the process of dissolving aliment in the stomach. Digest, to dissolve by the action of a solvent, to infuse any medical substance in spirits. Discutient, an application to disperse a tumor. Diuretic, a medicine that icreases the secretion of urine. Drastic, strong, active, violent.

Efflorescence, redness of the skin around an eruption. Emetic, a medicine which excites vomiting. Emmenagogue, that which promotes the flow of the menses. Emolient, that which softens and relaxes the solids. Epidemic, Page 306 a contagious disease, attacking many people the same season. Errhines, articles that excite sneezing. Eruption, breaking out on the skin. Excoriate, to strip off the skin. Excrement, the alvine, foeces, or stools.

Foeces, excrements or stools. Febrile, indicating fever, pertaining to fever. Febrifuge, that which removes fever. Fibrous, consisting of small threads. Flatulency, windiness in the stomach and intestines. Fomentation, the application of flannel dipped in hot water.--Flooding, an excessive flow of the menses. Friction, the act of rubbing. Fracture, a broken bone. Fundament, the aperture from which the excrements are ejected, the seat. Fungus, proud flesh, or any other excrescence. Fur, the coat of morbid matter upon the tongue.   Gangrene, the incipient or forming stage of mortification. Gargle, a wash for the mouth and threat.   Hectic, a slow fever. Hemorrhages, a discharge of bleed. Hemorrhoids, the piles. Hepatic, pertaining to the liver. Hydragogue, that which promotes the discharge of humors from the body. Hypocondriacal, low spirited. Hymen, the virginal membrane, partly closing the passage of the vagina. Hysterics, a disease peculiar to women, characterized by spasmodical and nervous affections, and often attended with hypocondriacal symptoms.

Idiophatic, a term applied to diseases that exist independent of all other complaints. Idiosyncrasy, the peculiar temperament or constitution of the body. Indented, notched. Indigenous, native. Infectious, communicating disease by contagion. Infuse, to steep in a liquid without boiling. Intestines, the tubes in the abdomen, vulgarly called guts. Intermittent, ceasing for intervals of time.   Jagged, uneven, having jaggs or teeth.

Lanceolate, oblong, shaped like a lancet. Laxatives, a gentle cathartic. Ligature, a bandage. Ligament, a strong membrane connecting the joints. Lithotomy, the operation of cutting the stone out of the bladder.

Materia Medica, description of medicine. Meconium, the first stools of an infant. Membrane, a thin delicate skin. Menses, monthly courses of females. Menstruation, the act of discharging the menses. Menstrual, pertaining to the menses. Miasm, putrid exhalations. Menstruum, any fluid used as a solvent. Morbid, diseased, unhealthy.

Mucilage, a glutinous, slimy substance. Mucous, the slimy fluid secreted by the mucous membrane. Muscles, the organs of motion.

Narcoctic, that which produces sleep by stupefaction. Nausea, inclination to vomit. Nervine, that which relieves disorders of the nerves. Oblong, longer than broad. Obtuse, a dull, heavy pain opposite to acute. Organ, any part capable of preparing some distinct operation. Orifice, an opening.

Paralytic, relating to palsy. Paroxysm, a periodical attack or fit of a disease. Peduncle, the stem that supports the flower. Perennial, in botany, a plant that lives more than two years. Perspiration, evacuation of fluid matter through the pores of the skin. Petioles, the foot stalks of a leaf. Pinnate, a compound leaf, composed of one stem and several small leaves on each side of it. Plethory, a fulness of habit fulness of the vessels. Pulmonary, pertaining to the lungs.

Quartan, recurring every fourth day. Quotidian, recurring every day.

Racemes, growing in clusters. Radiating, spreading or shooting in the form of rays. Radical, pertaining to the root. Rectum, that part of the intestines that reaches to the anus. Respiration, the act of breathing.--Retching, straining to vomit. Rigidity, stiffness. Rigor, a sense of chillness, with contraction of the skin. Rubefacient, an application that reddens the skin without blistering.

Saliva, the spittle. Secretion, the act of separating substances from the blood. Serrate, notched like a saw. Sinapism, a poultice of mustard, vinegar and flour. Solvent, that which has the power of dissolving. Stimulants, medicines that excite action and energy in the system. Stranguary, difficulty in voiding urine. Styptics, medicines that check the flow of blood.

Sudorifics, medicines that produce sensible perspiration. Syphilis, the venereal disease. Tent, a roll of lint placed in the opening of an ulcer Terminal, terminating, growing at the end of a stem. Tertian, a disease whose paroxysms return every other day. Tonics, medicines that increase the tone and strength of the system. Tumor, a swelling. Typhoid, resembling typhus, weak, low. Triennial, lasting three years.

Umbel, a flower resembling an umbrella. Umbeliferous, bearing umbels. Umbilical, pertaining to the naval. Ulcer, an ill conditioned, running sore. Urethra, the canal conveying the urine. Uterus, the womb.

Vagina, the canal leading to the womb, Ventilation. a free admission of air. Vermifuge, medicines that expel worms. Vertigo, giddiness of the head. Viscera, the entrails.

Whorls, flowers or leaves growing round the stem in a ring.

NOTE.--We have corrected innumerable errors that were in the copy from which this work is printed, and altered the grammatical construction of many sentences, and yet there are very many left, but we hope none that will mislead the common reader, and none but what may be easily understood and corrected by those who are more particular in such things. Our absence at various times during the execution of the work, prevented us from doing that justice that the work otherwise would have received.

JAMES M. EDNEY.   **April 12, 1850.**

The Cherokee Physician, or Indian Guide to Health

www.ingramcontent.com/pod-product-compliance
Lightning Source LLC
Chambersburg PA
CBHW080907170526
45158CB00008B/2024